T0349917

ASYMPTOTIC INTEGRATION AND STABILITY

For Ordinary, Functional and Discrete
Differential Equations of Fractional Order

Series on Complexity, Nonlinearity and Chaos

ISSN 2010-0019

Series Editor: Albert C.J. Luo
(Southern Illinois University Edwardsville, USA)

Aims and Scope

The books in this series will focus on the recent developments, findings and progress on fundamental theories and principles, analytical and symbolic approaches, computational techniques in nonlinear physical science and nonlinear mathematics.

Topics of interest in Complexity, Nonlinearity and Chaos include but not limited to:

· New findings and discoveries in nonlinear physics and mathematics,
· Complexity and mathematical structures in nonlinear physics,
· Nonlinear phenomena and observations in nature,
· Computational methods and simulations in complex systems,
· New theories, and principles and mathematical methods,
· Stability, bifurcation, chaos and fractals in nonlinear physical science.

Vol. 1 Ray and Wave Chaos in Ocean Acoustics: Chaos in Waveguides
        D. Makarov, S. Prants, A. Virovlyansky & G. Zaslavsky

Vol. 2 Applications of Lie Group Analysis in Geophysical Fluid Dynamics
        N. H. Ibragimov & R. N. Ibragimov

Vol. 3 Fractional Calculus: Models and Numerical Methods
        D. Baleanu, K. Diethelm & J. J. Trujillo

Vol. 4 Asymptotic Integration and Stability: For Ordinary, Functional
        and Discrete Differential Equations of Fractional Order
        D. Baleanu & O. G. Mustafa

Series on Complexity, Nonlinearity and Chaos – Vol. 4

ASYMPTOTIC INTEGRATION AND STABILITY

For Ordinary, Functional and Discrete
Differential Equations of Fractional Order

Dumitru Baleanu

Institute of Space Sciences, Romania & Çankaya University, Turkey

Octavian G. Mustafa

University of Craiova, Romania

World Scientific

NEW JERSEY · LONDON · SINGAPORE · BEIJING · SHANGHAI · HONG KONG · TAIPEI · CHENNAI

Published by

World Scientific Publishing Co. Pte. Ltd.
5 Toh Tuck Link, Singapore 596224
USA office: 27 Warren Street, Suite 401-402, Hackensack, NJ 07601
UK office: 57 Shelton Street, Covent Garden, London WC2H 9HE

Library of Congress Cataloging-in-Publication Data
Baleanu, D. (Dumitru)
   Asymptotic integration and stability : for ordinary, functional and discrete differential equations
of fractional order / Dumitru Baleanu, Çankaya University, Ankara, Turkey, Institute of Space
Sciences, Bucharest, Romania, Octavian G. Mustafa, University of Craiova, Romania.
      pages cm. -- (Series on complexity, nonlinearity and chaos ; volume 4)
   Includes bibliographical references and index.
   ISBN 978-9814641098 (hardcover : alk. paper)
   1. Fractional calculus. 2. Fractional differential equations. I. Mustafa, Octavian G. II. Title.
QA314.B35 2015
515'.83--dc23

                                                                        2014045570

British Library Cataloguing-in-Publication Data
A catalogue record for this book is available from the British Library.

Typeset by Stallion Press
Email: enquiries@stallionpress.com

Printed in Singapore

To our wives

Mihaela-Cristina and Dorina

Preface

The aim of the fractional calculus is to study the fractional order integral and derivative operator over real and complex domains as well as their applications. We recall that the tools of fractional calculus are as old as calculus itself.

Nowadays there are strong motivations (the unelucidated nature of the dark matter and dark energy, the difficult reconciliation of Einstein's General Relativity (GR) and Quantum Theory) to consider alternative theories that modify, extend or replace GR. We recall that some of these theories presume a higher dimensional space-time, and part of them predict violations of the physics fundamental principles: the Equivalence Principle and Lorentz symmetry could be broken, the fundamental constants could vary, the space could be anisotropic, and the physics could become nonlocal.

Fractional calculus becomes very powerful in the study of the anomalous social and physical behaviors, where scaling power law of fractional order appears universal as an empirical description of the complex phenomena.

The classical mathematical models, including nonlinear models, do not give adequate results in many cases where power law is clearly present.

During the last years, the asymptotic integration and the stability of fractional differential equations become an important research topic in the field of fractional calculus and its applications. Thus, a better understanding of these concepts represents one of the major tasks for researchers working on these fields and related topics.

The book contains eleven chapters and it is based mainly on the results reported by the authors during the last few years. The first chapter is about the differential operators of order $1+\alpha$ and their integral counterpart. The second chapter describes the existence and the uniqueness of solution for the differential equations of order α. Chapter three debates the position of the zeros, the Bihari inequality and the asymptotic behaviour of solutions for the differential equations of order α. Chapter four describes the asymptotic integration for the differential equations of order $1 + \alpha$. In chapter five we present the existence and the uniqueness of solutions for some delay differential equations within Caputo derivative. In chapter six we discuss the existence and the positive solutions for some delay fractional differential equations with generalized N term. The stability of a class of discrete fractional nonautonomous systems is shown in chapter seven. Mittag-Leffler stability theorem for fractional nonlinear systems with delay is the subject of the chapter eight. Chapter nine is concentrated on the Razumikhin stability theorem for some fractional systems with delay. Chapter ten deals with the controlability of some fractional evolution nonlocal impulsive quasilinear delay integro-differential systems. Finally, the book ends with the approximate controllability of Sobolev type nonlocal fractional stochastic dynamic systems in Hilbert spaces.

We would like to thank to all of our co-authors who helped us in writing this book by providing many interesting comments and remarks.

Finally, we deeply thank the editorial assistance of World Scientific Publishing Co., especially Ms. L.F. Kwong and Mr. Rajesh Babu.

Dumitru Baleanu
Çankaya University, Ankara, Turkey
Institute of Space Sciences, Bucharest, Romania

Octavian Mustafa
University of Craiova, Romania

Contents

Preface   vii

1. The Differential Operators of Order $1 + \alpha$ and Their
    Integral Counterparts   1

    1.1   The Gamma Function   2
    1.2   The Riemann-Liouville Derivative   3
    1.3   The Abel Computation   19
    1.4   The Operators. The Caputo Differential   21
    1.5   The Integral Representation of the Operators.
          The Half-line Case   23

2. Existence and Uniqueness of Solution for the Differential
    Equations of Order α 27

    2.1   A Lovelady-Martin Uniqueness Result for the
          Equation (2.2)   34
    2.2   A Nagumo-like Uniqueness Criterion
          for the Fractional Differential Equations
          with a Riemann-Liouville Derivative   39
    2.3   A Wintner-type Existence Interval for the
          Equation (2.2)   45

3. Position of the Zeros, the Bihari Inequality, and the
    Asymptotic Behavior of Solutions for the Differential
    Equations of Order α 51

    3.1   A Fite-type Length Criterion for Fractional
          Disconjugacy   51

3.2   The Bihari Inequality   55

3.3   Asymptotic Integration of the Differential Equations
of Orders 1 and α   61

3.4   The Bihari Asymptotic Integration
Theory of the Differential Equations of Second Order . . .   76

4.   Asymptotic Integration for the Differential Equations of
Order $1 + \alpha$ 79

4.1   An Asymptotic Integration Theory of Trench Type   79

4.2   Asymptotically Linear Solutions   90

4.3   A Bihari-Like Result   95

4.4   Convergent Solutions   104

4.5 L^p–Solutions of the Equation (4.3)   109

5.   Existence and Uniqueness of Solution for Some Delay
Differential Equations with Caputo Derivatives   121

6.   Existence of Positive Solutions for Some Delay Fractional
Differential Equations with a Generalized N–Term   129

6.1   The Existence Theorem   130

6.2   Existence and Uniqueness for the Solution   137

7.   Stability of a Class of Discrete Fractional
Nonautonomous Systems   139

7.0.1   Preliminaries   139

7.0.2   The Lyapunov Method for Discrete Fractional
Nonautonomous Systems   140

8.   Mittag-Leffler Stability of Fractional Nonlinear Systems
with Delay   145

9.   Razumikhin Stability for Fractional Systems in the
Presence of Delay   151

10.   Controllability of Some Fractional Evolution Nonlocal
Impulsive Quasilinear Delay Integro-Differential Systems   157

10.1   Preliminaries   157
10.2   The Problem   161
10.3   A Controllability Result   162

11.   Approximate Controllability of Sobolev Type Nonlocal
     Fractional Stochastic Dynamic Systems in Hilbert Spaces   169

     11.0.1   The Problem   170
     11.0.2   Approximate Controllability   175

Bibliography   181

Index   195

Chapter 1

The Differential Operators of Order $1 + \alpha$ and Their Integral Counterparts

When asked about the significance of the *non-integer order* for a differential operator by anyone without special interest in the mathematics of differential equations, the *fractionist*[1] can provide the inquirer with a simple analogy, described in the following.

Take a continuous function $f : I \to \mathbb{R}$, where $I = [a, b]$ is an interval of real numbers, and write down the identity below

$$f(t) = \frac{d}{dt} \left[\int_a^t f(s)ds \right]$$
$$= (\text{Diff} \circ \text{Int})(f)(t) = \mathcal{E}(f)(t), \quad t \in (a, b).$$

The identity expresses the fact that you did one integration — the order of the expression $\mathcal{E}(f) = \text{Int}(f)$ is now +1 — followed by one differentiation — the new, and final, order of $\mathcal{E}$ is $(+1) + (-1) = 0$ —.

Let's go further and recall, via integration by parts, that

$$f(t) = \frac{1}{(n-1)!} \cdot \frac{d^n}{dt^n} \left[\int_a^t (t-s)^{n-1} f(s)ds \right], \quad t \in (a, b). \qquad (1.1)$$

As before, the integral brings +1 into the sum (for computing the order), while the "$t - s$" component is responsible for another $n - 1$ *units* of order. The final order of $\mathcal{E}$ is $(+1) + (n-1) \cdot (+1) + n \cdot (-1) = 0$.

Now, given $z \in \mathbb{C}$ and t, $\varepsilon > 0$, the mapping $z \mapsto t^z = e^{z \cdot \log t}$ is entire and it makes sense to wonder about the function $z \mapsto \int_0^{t+\varepsilon} (t-s)^z f(s)ds$ being holomorphic [Hille (1959), pp. 72, 230], a property sometimes called upon as analytic [Ablowitz and Fokas (2003), p. 24]. Using the technique from e.g., [Rudin (1987), Chap. 10, Ex. 16], it can be established that the latter function is, in fact, entire. In particular, taking $\alpha \in \mathbb{R} \backslash \mathbb{N}$ and $n \in \mathbb{N}$,

---

[1]That is, the analyst of the "fractionals".

with $n \geq 1$, the quantity

$$\mathcal{E}(f)(t) = \frac{d^n}{dt^n}\left[\int_a^{t+\varepsilon}(t-s)^\alpha f(s)ds\right], \quad t \in (a,b),$$

makes sense. Without paying any attention[2] to the extra "ε", deduce that the order of the expression must be $(+1) + \alpha \cdot (+1) + n \cdot (-1) = \alpha - n + 1$.

To conclude our analogy, we get rid of ε by making it tend to zero and look for some positive constant to mimic the $\frac{1}{(n-1)!}$ coefficient from (1.1). The formula of a new derivative having a *non-integer* order "close" to n is expressed as

$$c_n \cdot \frac{d^n}{dt^n}\left[\int_a^t(t-s)^{\alpha_n}f(s)ds\right], \quad t \in (a,b), \tag{1.2}$$

where $\alpha_n \in (-1,0]$ and $c_n > 0$. It resembles the classical Riemann-Liouville construction [Podlubny (1999a), p. 68].

1.1   The Gamma Function

For a proper constant c_n, we rely on Euler's integral (of second kind) Γ. This function can be stated as

$$\Gamma(z) = \int_0^{+\infty}e^{-t}t^{z-1}dt, \quad \text{where } t^\xi = \begin{cases} 0, & t = 0, \\ e^{\xi \cdot \log t}, & t > 0, \end{cases} \quad \xi, z \in \mathbb{C},$$

and the real part of z is positive [Olver (1974), p. 31].

Since we shall employ everywhere only the most common of its properties, we remind the interested reader that he or she can delve anytime into the detailed proofs of results about the Gamma function from [Olver (1974); Podlubny (1999a)]. As for ourselves, just remember that

$$\begin{aligned}
&\Gamma(z+1) = z \cdot \Gamma(z), \\
&\Gamma(n) = (n-1)!, && n \in \mathbb{N}\backslash\{0\}, \\
&\Gamma(z) \cdot \Gamma(1-z) = \frac{\pi}{\sin \pi z}, && \text{when } z \text{ is non-integer,} \\
&\frac{\Gamma(z_1) \cdot \Gamma(z_2)}{\Gamma(z_1+z_2)} = \int_0^1 v^{z_1-1}(1-v)^{z_2-1}dv, \, z_1, z_2 \in \mathbb{C},
\end{aligned} \tag{1.3}$$

where the real parts of z_1, z_2 are positive. The last quantity in (1.3) is the Euler integral of first kind, also referred to as the Beta function $B(z_1, z_2)$ [Olver (1974), p. 37].

Following [Baleanu et al. (2011c)], we claim as well that

$$\Gamma(\beta) > \frac{1}{\Gamma(2-\beta)}, \quad \beta \in (0,1). \tag{1.4}$$

---

[2]For a technical approach, see [Stein (1970), p. 77, Lemma].

To prove the assertion, just remark that

$$\Gamma(\beta) \cdot \Gamma(2 - \beta) = (1 - \beta)\Gamma(1 - \beta) \cdot \Gamma(\beta) = (1 - \beta) \cdot \frac{\pi}{\sin(\pi\beta)}$$

$$= \frac{\pi(1 - \beta)}{\sin(\pi(1 - \beta))},$$

which leads to

$$\lim_{\beta \searrow 0} \Gamma(\beta) \cdot \Gamma(2 - \beta) = +\infty, \quad \lim_{\beta \nearrow 1} \Gamma(\beta) \cdot \Gamma(2 - \beta) = 1.$$

Observe also that there are no critical points of the function $\beta \mapsto \frac{\pi(1-\beta)}{\sin(\pi\beta)}$ in $(0, 1)$. This follows from the fact that the unique solution of the transcendental equation $\tan \pi(1 - \beta) = \pi(1 - \beta)$ in $[0, 1]$ is $\beta^\star = 1$. So,

$$\Gamma(\beta) \cdot \Gamma(2 - \beta) > 1, \quad \beta \in (0, 1).$$

The claim is established.

1.2   The Riemann-Liouville Derivative

A multitude of notations has been designed to capture the complexity of fractional differentiation and integration, see the monographs [Samko et al. (1993); Miller and Ross (1993); Kilbas et al. (2006)] or the classical treatise [Hille and Phillips (1957), pp. 664, 673]. In a simplified version, the *Riemann-Liouville derivative*, of *order* $\alpha \in (0, 1)$, reads as below

$$_0D_t^\alpha(f)(t) = \frac{1}{\Gamma(1 - \alpha)} \cdot \frac{d}{dt}\left[\int_0^t \frac{f(s)}{(t - s)^\alpha}ds\right], \quad t > 0. \tag{1.5}$$

The subscripts 0 and t in this symbol of derivative[3] hint at the integration interval $(0, t)$, recall (1.2).

The first issue regarding (1.5) is about the existence of the integral inside. As the mapping $\eta_{t,\alpha}$ given by $s \mapsto (t-s)^{-\alpha}$ is a member of $L^1((0, t), \mathbb{R})$, it is natural to ask that $f \in L^\infty((0, t), \mathbb{R})$, see [Rudin (1987), Chap. 3, Th. 3.8]. In particular, this qualifies all the functions from $C([0, t], \mathbb{R})$ as candidates for f.

Since any absolutely integrable function — with respect to the Lebesgue measure on $[0, t]$ — may take infinite values on a null-measure set, we see that $f\eta_{t,\alpha}$ might be infinite in other points beside $s = t$. To take advantage of this remark, introduce $\mathcal{RL}^\beta = \mathcal{RL}^\beta((0, +\infty), \mathbb{R})$ the real linear space of

---

[3]Given the linearity of the right-hand part of (1.5), we shall use freely either of symbols $_0D_t^\alpha(f)$, $_0D_t^\alpha f$ when referring to the Riemann-Liouville derivative of f.

all the functions $f \in C((0, +\infty), \mathbb{R})$ with $\lim_{t \searrow 0}[t^\beta f(t)] \in \mathbb{R}$ for some $\beta \in [0, 1)$.

Now,

$$\int_0^t \frac{f(s)}{(t-s)^\alpha} ds = \int_0^t \frac{s^\beta \cdot f(s)}{s^\beta(t-s)^\alpha} ds \tag{1.6}$$

$$= \left[\int_0^{+\infty} \frac{\chi_{(0,\frac{t}{2})}}{s^\beta(t-s)^\alpha} + \int_0^{+\infty} \frac{\chi_{(\frac{t}{2},t)}}{s^\beta(t-s)^\alpha} \right] \cdot s^\beta f(s) \, ds.$$

Again, the ratios inside the integrals are L^1-functions while $s \mapsto s^\beta f(s)$ is in[4] $L^\infty_{\text{loc}}([0, +\infty), \mathbb{R})$ for any $f \in \mathcal{RL}^\beta$. Here, as usual, χ designates the characteristic function of a Lebesgue-measurable set. From now on we shall presume that *every function f involved in a Riemann-Liouville differentiation is a member of some $\mathcal{RL}^\beta$*.

The second issue concerning (1.5) is the differentiability of the integral. To find some reasonable restrictions on f that will lead to this differentiability, notice that we can perform the change of variables $s = t \cdot v$ for the integral inside (1.5), that is

$$\int_0^t \frac{f(s)}{(t-s)^\alpha} ds = t^{1-\alpha} \cdot \int_0^1 \frac{f(tv)}{(1-v)^\alpha} dv, \tag{1.7}$$

see [Rudin (1987), Chap. 7, Th. 7.26].

Assume now that the function f from (1.7) is (locally) absolutely continuous in $[0, +\infty)$, which means it is differentiable *almost everywhere* — we shall use the shorthand notation **a.e.** — and $f' \in L^1_{\text{loc}}([0, +\infty), \mathbb{R})$ [Rudin (1987), Chap. 7, Th. 7.20]. Moreover, we ask that the mapping $s \mapsto s^{1+\beta} f'(s)$ be in $L^\infty_{\text{loc}}([0, +\infty), \mathbb{R})$ for some $\beta \in [0, 1)$. We shall refer to this mapping as $s^{1+\beta} f'$ in the following computation. A significant particular case of our restriction is when $f' \in L^\infty_{\text{loc}}([0, +\infty), \mathbb{R})$, which makes f a *locally Lipschitz function*. We recall Rademacher's theorem [Evans and Gariepy (1992), p. 81] which deals with the a.e. differentiability of such functions. Another important situation is when $f \in C^1((0, +\infty), \mathbb{R}) \cap \mathcal{RL}^\gamma$ and $f' \in \mathcal{RL}^{1+\beta}$ for some $\gamma, \beta \in [0, 1)$. These restrictions have been modeled to ensure the existence of the right-hand part of (1.11).

Allow us to recapitulate at this point several elementary facts. First, given $\beta \in (0, 1)$ and $x \in (0, 1]$, since $\ln x \leq 0$, we have $\beta \ln x \geq \ln x$ and, by exponentiation, $x^\beta \geq x$. Second, given $\alpha \in (0, 1)$, we get

$$1 = (1-\alpha) + \alpha \leq (1-\alpha)^\beta + \alpha^\beta.$$

---

[4]Recall that, as a consequence of Luzin's theorem [Rudin (1987), Chap. 2, Th. 2.24], a function $f \in L^\infty([a, b], \mathbb{R})$ is, almost everywhere with respect to the Lebesgue measure on $[a, b]$, the pointwise limit of a sequence of compactly supported continuous functions. See also [Rudin (1987), Chap. 9, Sect. 9.22].

Third, given A, $B > 0$, take $\alpha = \frac{A}{A+B}$. Then, the previous inequality yields

$$(A + B)^\beta \leq A^\beta + B^\beta. \tag{1.8}$$

It is obvious that the inequality remains valid whenever A, $B \geq 0$ and $\beta \in (0, 1)$. As a by-product, when $A > 0$ and $B \in [0, 1)$, we have

$$\frac{1}{A^{1-\beta}(A+B)^\beta} = \frac{1}{A} \cdot \left(\frac{A}{A+B} \right)^\beta \geq \frac{1}{A} \cdot \frac{A}{A+B} \geq \frac{1}{1+A}. \tag{1.9}$$

The previous restrictions on f imply that, given $h > 0$ small, the function g_h with the formula — the number $t > 0$ is fixed —

$$g_h(v) = \frac{f((t+h)v) - f(tv)}{h} - vf'(tv), \quad \text{a.e. } v \in [0, 1],$$

belongs to $L^1([0, 1], \mathbb{R})$. In fact[5], after setting $\varepsilon > h > 0$,

$$|g_h(v)| \leq \frac{1}{h} \int_{tv}^{(t+h)v} |f'(s)| ds + v|f'(tv)|$$

$$\leq \|s^{1+\beta} f'\|_{L^\infty([tv,(t+h)v], \mathbb{R})} \cdot \frac{1}{h} \int_{tv}^{(t+h)v} \frac{ds}{s^{1+\beta}} + v|f'(tv)|$$

$$= \|s^{1+\beta} f'\|_{L^\infty([tv,(t+h)v], \mathbb{R})} \frac{1}{\beta h v^\beta} \left[\frac{1}{t^\beta} - \frac{1}{(t+h)^\beta} \right] + v|f'(tv)|$$

$$= \|s^{1+\beta} f'\|_{L^\infty([tv,(t+h)v], \mathbb{R})} \frac{1}{\beta h v^\beta} \cdot \frac{\beta h}{\xi^{1+\beta}} + v|f'(tv)| \tag{1.10}$$

$$\leq \|s^{1+\beta} f'\|_{L^\infty([tv,(t+h)v], \mathbb{R})} \frac{1}{v^\beta} \cdot \frac{1}{t^{1+\beta}} + \frac{(tv)^{1+\beta}}{t^{1+\beta} v^\beta} \cdot |f'(tv)|$$

$$\leq \frac{2}{t^{1+\beta}} \|s^{1+\beta} f'\|_{L^\infty([0,t+\varepsilon], \mathbb{R})} \cdot \frac{1}{v^\beta} = H_{t,\beta}(v).$$

We have employed the mean value theorem in (1.10), so, $\xi \in (t, t+h)$. If we had used (1.8) instead then we would have got the quantity $\frac{1}{h} \cdot \left(\frac{h}{t+h} \right)^\beta$ as a replacement of $\frac{\beta}{\xi^{1+\beta}} \cdot h$. The former quantity, however, is bounded from below only, see (1.9).

Since the functions $v \mapsto \frac{g_h(v)}{(1-v)^\alpha}$, where $h \in (0, \varepsilon)$, converge pointwisely to zero almost everywhere in $[0, 1]$ when $h \searrow 0$ and they are all bounded in absolute value by the integrable function $v \mapsto \frac{H_{t,\beta}(v)}{(1-v)^\alpha}$, the Lebesgue dominated convergence theorem establishes that their convergence is an

---

[5] Remark that the mapping $t \mapsto \frac{1}{t} \cdot \chi_{(0,+\infty)}(t)$ is in $L^\infty_{\text{loc}}((0, +\infty), \mathbb{R})$ but not in $L^\infty_{\text{loc}}([0, +\infty), \mathbb{R})$ while the mapping $t \mapsto t \sin \frac{1}{t^2} \cdot \chi_{(0,+\infty)}(t)$ belongs to both function spaces.

L^1-convergence over $[0,1]$, see [Evans and Gariepy (1992), p. 20]. We have obtained

$$\frac{d}{dt}\left[\int_0^1 \frac{f(tv)}{(1-v)^\alpha}dv\right] = \int_0^1 \frac{vf'(tv)}{(1-v)^\alpha}dv = t^{\alpha-2}\int_0^t \frac{sf'(s)}{(t-s)^\alpha}ds, \quad t > 0.$$

Returning to (1.7), we deduce that

$$\Gamma(1-\alpha)\cdot {}_0D_t^\alpha(f)(t) = \frac{1-\alpha}{t}\int_0^t \frac{f(s)}{(t-s)^\alpha}ds + \frac{1}{t}\int_0^t \frac{sf'(s)}{(t-s)^\alpha}ds, \quad (1.11)$$

see [Baleanu et al. (2011a), p. 8].

As before, one might ask whether the restriction regarding $s^{1+\beta}f'$ is essential or just a trick to help shortening the proof. In fact, looking closely to the right-hand part of (1.11), we would like to know the significance of the mapping $\mathfrak{g} = \mathfrak{g}_\alpha$ given by $t \mapsto \int_0^t \frac{g(s)}{(t-s)^\alpha}ds$ when $g \in L^1_{\text{loc}}([0,+\infty),\mathbb{R})$. We claim that *the previous mapping is a member of* $L^1_{\text{loc}}([0,+\infty),\mathbb{R})$.

To establish the validity of this claim, set $t > 0$ and introduce the functions $F = F_\alpha$, $G = G_g$ with $F(x) = x^{-\alpha}\cdot \chi_{(0,\frac{t}{2})}(x)$ and $G(x) = |g(x)|\cdot \chi_{(0,t)}(x)$ for every $x \in \mathbb{R}$. Obviously, $F, G \in L^1(\mathbb{R},\mathbb{R})$. Computing their convolution [Rudin (1987), Chap. 8, Th. 8.14], we get

$$(F \star G)(x)$$

$$= \int_{-\infty}^{+\infty} F(x-y)G(y)dy = \begin{cases} 0, & x \le 0, \\ \int_0^x \frac{|g(y)|}{(x-y)^\alpha}dy, & 0 < x \le \frac{t}{2}, \\ \int_{x-\frac{t}{2}}^x \frac{|g(y)|}{(x-y)^\alpha}dy, & \frac{t}{2} < x \le t, \\ \int_{x-\frac{t}{2}}^t \frac{|g(y)|}{(x-y)^\alpha}dy, & t < x \le \frac{3t}{2}, \\ 0, & x > \frac{3t}{2}. \end{cases} \quad (1.12)$$

Since $F\star G \in L^1(\mathbb{R},\mathbb{R})$, we deduce that $(F\star G)\cdot\chi_{(0,\frac{t}{2})}$ is Lebesgue integrable. Now, as $|\mathfrak{g}(x)| \le \int_0^x \frac{|g(y)|}{(x-y)^\alpha}dy$ everywhere in $\left[0,\frac{t}{2}\right]$, the correctness of the claim is proved.

Allow us to take the reasoning one step further. Set $\alpha, \beta \in [0,1)$ and consider $g \in L^1_{\text{loc}}([0,+\infty),\mathbb{R})$. Now, as $\mathfrak{g}_\alpha \in L^1_{\text{loc}}([0,+\infty),\mathbb{R})$, let $F = F_\beta$ and $G = G_{\mathfrak{g}_\alpha}$ in the previous computation. We conclude that the mapping $t \mapsto \int_0^t \frac{1}{(t-s)^\beta}\int_0^s \frac{g(\tau)}{(s-\tau)^\alpha}d\tau ds$ is in $L^1_{\text{loc}}([0,+\infty),\mathbb{R})$. In particular, it is a.e. finite, and so

$$\int_0^t \frac{1}{(t-s)^\beta}\int_0^s \frac{|g(\tau)|}{(s-\tau)^\alpha}d\tau ds < +\infty, \quad \text{a.e. } t > 0. \quad (1.13)$$

The powerful device of convolution is not needed, of course, when we employ some "nice" functions. To see this, let us return to (1.11) and assume that f is absolutely continuous locally in $[0, +\infty)$ and $s^\beta f' \in L^\infty_{\text{loc}}([0, +\infty),$ $\mathbb{R})$. We have to recall several facts. First, an a.e. pointwise limit of (Lebesgue) measurable functions is also measurable [Rudin (1987), Chap. 1, Th. 1.14]. Second, all the elements of $L^1(I, \mathbb{R})$, where $I \subset \mathbb{R}$ is a compact interval, can be approximated[6] with continuous functions (a consequence of Luzin's theorem) [Rudin (1987), Chap. 3, Th. 3.14]. Third, a less known theorem of Lebesgue (1905, cf. [Rudin (1987), p. 384]) states that any function of two variables which is (separately) continuous in each variable is Borel measurable in the plane [Rudin (1987), Chap. 8, Ex. 8(b)]. In the same spirit, by taking $g(s, \tau) = s - \tau$ and $f(x) = x^{-\alpha} \chi_{(0, +\infty)}(x)$ in [Rudin (1987), Chap. 1, Th. 1. 12(d)] and noticing that they are Borel measurable, we deduce that the mapping $(s, \tau) \mapsto (s - \tau)^{-\alpha} \chi_{(0, +\infty)}(s - \tau) = (f \circ g)(s, \tau)$ is Borel measurable in the plane and nonnegative-valued. The same reasoning applies to the function $(s, \tau) \mapsto |s - \tau|^{-\alpha} \cdot \chi_{\mathbb{R}-\{0\}}(s - \tau)$. Fourth, the Lebesgue two-dimensional measure is a completion of the product of two one-dimensional Lebesgue measures [Rudin (1987), Chap. 8, Th. 8.11].

Now, the mapping $(s, \tau) \mapsto \frac{f'(\tau)}{|t-s|^\alpha} \cdot \chi_{\mathbb{R}-\{0\}}(t - s)$, where $t > 0$ is fixed, is an a.e. pointwise limit of products of continuous approximations of the (L^1) factors $\tau \mapsto f'(\tau)$ and $s \mapsto |t - s|^{-\alpha} \chi_{\mathbb{R}-\{0\}}(t - s)$. These products are Borel measurable in the plane — the third fact — and, as a consequence, they are Lebesgue measurable in the plane [Rudin (1987), Chap. 2, Th. 2.20]. By being an a.e. (pointwise) limit, the mapping itself is Lebesgue measurable in the plane. The same reasoning applies, obviously, to its absolute value $(s, \tau) \mapsto \frac{|f'(\tau)|}{|t-s|^\alpha}$. We multiply the latter mapping with $(s, \tau) \mapsto |s - \tau|^{-\gamma} \chi_{\mathbb{R}-\{0\}}(s - \tau)$ for some $\gamma \in [0, 1)$. Our intention, in the following, is to use the Fubini theorem [Rudin (1987), Chap. 8, Th. 8.8] to manipulate the (bi-dimensional Lebesgue) measurable function

$$(s, \tau) \mapsto \frac{|f'(\tau)|}{|t - s|^\alpha |s - \tau|^\gamma} \cdot \chi_{\mathbb{R}-\{0\}}(t - s) \chi_{\mathbb{R}-\{0\}}(s - \tau). \qquad (1.14)$$

We need an auxiliary result: *the function $(s, \tau) \mapsto |t - s|^{-\alpha} s^{-\beta} |s - \tau|^{-\gamma}$, where $\alpha, \beta, \gamma \in [0, 1)$ and $t, \tau > 0$ with $t \neq \tau$ are fixed, is Lebesgue measurable in the plane and its "partial integrals" are a.e. finite.* We shall deal in the sequel only with the second part of the statement. According to

---

[6] The cited result ensures an approximation in the L^1-norm. However, if a sequence of integrable functions convergences in this norm, the sequence possesses a sub-sequence that is a.e. pointwisely convergent to the L^1-limit of the sequence [Rudin (1987), Chap. 3, Th. 3.12]. So, *any integrable function is a.e. pointwisely approximated by a sequence of continuous functions.*

"part (b)" of Fubini's theorem[7], to establish that the simpler application $(s, \tau) \mapsto s^{-\beta} |s - \tau|^{-\gamma}$, where $\beta, \gamma \in [0, 1)$, is (Lebesgue) integrable in $[0, a] \times [0, b]$ for some $a, b > 0$, it suffices prove that its integral $\mathcal{I}_s$ with respect to τ and the integral of $\mathcal{I}_s$ with respect to s are both a.e. finite. This is the conclusion of (1.15), (1.17) from the simple estimates below

$$\int_0^u \frac{d\tau}{s^\beta |s - \tau|^\gamma}$$

$$\leq \frac{1}{s^\beta} \left(\int_0^s + \chi_{(0,+\infty)}(u - s) \cdot \int_s^u \right) \frac{d\tau}{|s - \tau|^\gamma}$$

$$\leq \frac{1}{(1 - \gamma) s^\beta} \cdot \left[s^{1-\gamma} + |u - s|^{1-\gamma} \right] < +\infty, \quad u \in [0, b], \qquad (1.15)$$

where $s > 0$, and

$$\int_0^v \frac{ds}{s^\beta |s - \tau|^\gamma}$$

$$\leq \left(\int_0^{\frac{\tau}{2}} + \chi_{(0,+\infty)} \left(v - \frac{\tau}{2} \right) \cdot \int_{\frac{\tau}{2}}^v \right) \frac{ds}{s^\beta |s - \tau|^\gamma}$$

$$\leq \left(\frac{2}{\tau} \right)^\gamma \cdot \int_0^{\frac{\tau}{2}} \frac{ds}{s^\beta} + \left(\frac{2}{\tau} \right)^\beta \cdot \chi_{(0,+\infty)} \left(v - \frac{\tau}{2} \right) \int_{\frac{\tau}{2}}^v \frac{ds}{|s - \tau|^\gamma}$$

$$\leq \frac{1}{1 - \beta} \left(\frac{\tau}{2} \right)^{1-\gamma-\beta} + \left(\frac{2}{\tau} \right)^\beta \int_0^v \frac{ds}{|s - \tau|^\gamma}$$

$$\leq \frac{1}{1 - \beta} \left(\frac{\tau}{2} \right)^{1-\gamma-\beta} + \left(\frac{\tau}{2} \right)^{-\beta} \cdot \frac{\tau^{1-\gamma} + |v - \tau|^{1-\gamma}}{1 - \gamma}, \quad v \in [0, a], \quad (1.16)$$

where $\tau > 0$, and

$$\int_0^a \mathcal{I}_s ds$$

$$= \int_0^a \left[\int_0^b \frac{d\tau}{s^\beta |s - \tau|^\gamma} \right] ds \leq \frac{1}{1 - \gamma} \int_0^a \frac{s^{1-\gamma} + |b - s|^{1-\gamma}}{s^\beta} ds$$

$$\leq \frac{a^{1-\gamma} + (|b| + |a|)^{1-\gamma}}{1 - \gamma} \cdot \int_0^a \frac{ds}{s^\beta}$$

$$= \frac{a^{2-\gamma-\beta} + (|b| + |a|)^{1-\gamma} a^{1-\beta}}{(1 - \gamma)(1 - \beta)} < +\infty. \qquad (1.17)$$

Further, let's apply the "part (b)" to the function from the statement of the auxiliary result. Via (1.15), we have

$$\int_0^u \frac{d\tau}{|t - s|^\alpha s^\beta |s - \tau|^\gamma} \leq \frac{\mathcal{I}_s}{|t - s|^\alpha}$$

---

[7] As presented in Rudin's treatise [Rudin (1987)].

$$\leq \mathcal{J}_s = \frac{a^{1-\gamma} + (|b| + |a|)^{1-\gamma}}{(1-\gamma)s^\beta |t-s|^\alpha}, \quad u \in [0, b],$$

where $t \neq s$ and $s > 0$, and

$$\int_0^a \mathcal{J}_s ds \leq \frac{a^{1-\gamma} + (|b| + |a|)^{1-\gamma}}{1-\gamma} \cdot \int_0^a \frac{ds}{s^\beta |s-t|^\alpha}$$

$$\leq \frac{a^{1-\gamma} + (|b| + |a|)^{1-\gamma}}{1-\gamma}$$

$$\times \left[\frac{1}{1-\beta} \left(\frac{t}{2}\right)^{1-\alpha-\beta} + \left(\frac{t}{2}\right)^{-\beta} \frac{t^{1-\alpha} + |a-t|^{1-\alpha}}{1-\alpha} \right] < +\infty,$$

where $t > 0$. We have employed (1.16) with $\tau = t$ and $\gamma = \alpha$. According to Fubini's theorem, the function in the auxiliary result belongs to $L^1_{\text{loc}}([0, +\infty)^2, \mathbb{R})$. The conclusion[8] of the result is reached via "part (c)" of the Fubini theorem.

Now, since $s^\beta f' \in L^\infty_{\text{loc}}([0, +\infty), \mathbb{R})$, we get

$$\frac{|f'(\tau)|}{|t-s|^\alpha |s-\tau|^\gamma} \leq \frac{\|\tau^\beta f'\|_{L^\infty([0,b],\mathbb{R})}}{|t-s|^\alpha s^\beta |s-\tau|^\gamma}, \quad (s, \tau) \in [0, a] \times [0, b], \quad (1.18)$$

and so *the mapping* $(s, \tau) \mapsto \frac{f'(\tau)}{|t-s|^\alpha |s-\tau|^\gamma} \cdot \chi_{\mathbb{R}-\{0\}}(t-s)\chi_{\mathbb{R}-\{0\}}(s-\tau)$ *is a member of* $L^1_{\text{loc}}([0, +\infty)^2, \mathbb{R})$. In particular, recalling that $\chi_{A \cap B} = \chi_A \cdot \chi_B$ when $A, B \subseteq \mathbb{R}$, the application $(s, \tau) \mapsto \frac{f'(\tau)}{|t-s|^\alpha |s-\tau|^\gamma} \chi_{(0,+\infty)}(t-s)\chi_{(0,+\infty)}(s-\tau)$ is in $L^1_{\text{loc}}([0, +\infty)^2, \mathbb{R})$. Its L^1–norm in $[0, a] \times [0, b]$ is exactly the integral (1.13) for $g = f'$. So, to answer to the (presumptive) question from page 6, *yes*, it was a trick!

The formula (1.11) can be connected with the key estimate in [Samko et al. (1993), p. 35, Lemma 2.2] from which we deduce that, *if f is absolutely continuous in $[0, +\infty)$ and*[9] $s^\beta f' \in L^\infty_{\text{loc}}([0, +\infty), \mathbb{R})$ *for some* $\beta \in [0, 1)$, *the Riemann-Liouville derivative can be recast as*

$$\Gamma(1-\alpha) \cdot {}_0D_t^\alpha(f)(t) = \frac{f(0)}{t^\alpha} + \int_0^t \frac{f'(s)}{(t-s)^\alpha} ds, \quad \text{a.e. } t > 0. \quad (1.19)$$

To see this, let us start by claiming that, *for any $g \in \mathcal{RL}^\beta$, where* $\beta \in (0, 1)$,

$$\frac{d}{dt}\left[\int_0^t (t-s)^\alpha g(s) ds \right] = \alpha \int_0^t \frac{g(s)}{(t-s)^{1-\alpha}} ds, \quad t > 0, \quad (1.20)$$

---

[8]It remains to demonstrate only that $\int_0^v \frac{ds}{|t-s|^\alpha s^\beta |s-\tau|^\gamma}$, where $t \neq \tau > 0$, is a.e. finite in $[0, a]$.

[9]Again, this hypothesis is inserted just to simplify the computations.

see [Baleanu et al. (2011a), p. 6].

A proof of our claim can be modeled following the lines preceding (1.11). Let's give, however, a slightly different approach. By denoting with $\mathcal{I}(t)$ the integral from the left-hand part of (1.20) and taking $t > 2h > 0$, we have

$$
\frac{\mathcal{I}(t+h) - \mathcal{I}(t)}{h} - \alpha \int_0^t \frac{g(\tau)}{(t-\tau)^{1-\alpha}} d\tau
$$

$$
= \int_0^{t-2h} \left[\frac{(t+h-\tau)^\alpha - (t-\tau)^\alpha}{h} - \frac{\alpha}{(t-\tau)^{1-\alpha}} \right] g(\tau) d\tau
$$

$$
+ \left\{ \int_{t-2h}^{t+h} \frac{(t+h-\tau)^\alpha}{h} g d\tau - \int_{t-2h}^t \left[\frac{(t-\tau)^\alpha}{h} + \frac{\alpha}{(t-\tau)^{1-\alpha}} \right] g d\tau \right\}
$$

$$
= \mathcal{I}_1(h) + \mathcal{I}_2(h). \tag{1.21}
$$

The sum of integrals in braces is estimated by

$$
|\mathcal{I}_2(h)|
$$

$$
\leq \int_{t-2h}^{t+h} \frac{(t+h-\tau)^\alpha}{h} |g| d\tau + \int_{t-2h}^t \left[\frac{(t-\tau)^\alpha}{h} + \frac{\alpha}{(t-\tau)^{1-\alpha}} \right] |g| d\tau
$$

$$
\leq \frac{(3h)^\alpha}{h} \int_{t-2h}^{t+h} d\tau \cdot \|g\|_{L^\infty([t-2h,t+h],\mathbb{R})} + \frac{(2h)^\alpha}{h} \int_{t-2h}^t d\tau \cdot \|g\|_{L^\infty([t-2h,t+h],\mathbb{R})}
$$

$$
+ \alpha \int_{t-2h}^t \frac{d\tau}{(t-\tau)^{1-\alpha}} \cdot \|g\|_{L^\infty([t-2h,t+h],\mathbb{R})}
$$

$$
= 3(2^\alpha + 3^\alpha) \cdot h^\alpha \cdot \|g\|_{L^\infty([t-2h,t+h],\mathbb{R})} = o(1) \quad \text{when } h \searrow 0.
$$

Recall now the Abel series, that is, the power series expansion of the real analytic function $v \mapsto (1+v)^\alpha$:

$$
(1+v)^\alpha = 1 + \alpha v + \frac{\alpha(\alpha-1)}{2} v^2 + \cdots = 1 + \alpha v + O(v^2), \quad |v| < 1,
$$

see [Ahlfors (1979), p. 180]. By taking $|v| \leq \frac{1}{2}$, there exists a constant $C = C(\alpha) > 0$ such that

$$
|(1+v)^\alpha - 1 - \alpha v| \leq C v^2, \quad |v| \leq \frac{1}{2}.
$$

Consequently, we get the next estimates for the first sum in (1.21)

$$
|\mathcal{I}_1(h)|
$$

$$
\leq \int_0^{t-2h} \left| \frac{(t+h-\tau)^\alpha - (t-\tau)^\alpha}{h} - \frac{\alpha}{(t-\tau)^{1-\alpha}} \right| \cdot |g| d\tau
$$

$$
\leq \int_0^{t-2h} \frac{(t-\tau)^\alpha}{h} \left| \left(1 + \frac{h}{t-\tau} \right)^\alpha - 1 - \alpha \cdot \frac{h}{t-\tau} \right| \cdot |g| d\tau
$$

$$\leq C \int_0^{t-2h} \frac{h}{(t-\tau)^{2-\alpha}} |g| d\tau \leq C \int_0^{t-2h} \frac{h}{(2h)^{1-\varepsilon}} \cdot \frac{|g(\tau)|}{(t-\tau)^{1-(\alpha-\varepsilon)}} d\tau$$

$$\leq 2^{\varepsilon-1} C \cdot h^\varepsilon \cdot \int_0^t \frac{\tau^\beta \cdot |g(\tau)|}{\tau^\beta (t-\tau)^{1-(\alpha-\varepsilon)}} d\tau$$

$$\leq 2^{\varepsilon-1} C h^\varepsilon \cdot \|\tau^\beta g\|_{L^\infty([0,t],\mathbb{R})} \int_0^1 \frac{dv}{v^\beta (1-v)^{1-(\alpha-\varepsilon)}} \cdot t^{\alpha-(\varepsilon+\beta)}$$

$$= 2^{\varepsilon-1} C h^\varepsilon \|\tau^\beta g\|_{L^\infty([0,t],\mathbb{R})} B(1-\beta, \alpha-\varepsilon) \cdot t^{\alpha-(\varepsilon+\beta)}$$

$$= O(h^\varepsilon) = o(1) \quad \text{when } h \searrow 0,$$

where $\varepsilon \in (0,\alpha)$. The claim in (1.20) is established.

Returning to (1.19), assume that $f \in C^1([0,T],\mathbb{R})$ for some $T > 0$. Now, via the classical integration by parts, we have

$$\int_0^t \frac{f(s)}{(t-s)^\alpha} ds = \lim_{\varepsilon \searrow 0} \int_\varepsilon^t \left[-\frac{(t-s)^{1-\alpha}}{1-\alpha} \right]' f(s) ds$$

$$= \frac{1}{1-\alpha} \left[f(0) t^{1-\alpha} + \int_0^t (t-s)^{1-\alpha} f'(s) ds \right].$$

To differentiate the latter identity, we apply (1.20) for $g = f'$, and this yields (1.19).

So, (1.19) works for convenient functions f. *But is it the same thing as (1.11)?* To answer affirmatively to this question, notice that the right-hand part of (1.19) is equal to

$$\frac{f(0)}{t^\alpha} + \frac{1}{t} \int_0^t \frac{t f'(s)}{(t-s)^\alpha} ds.$$

Subtracting this quantity from the right-hand part of (1.11), we observe that our task reduces to proving that

$$\frac{1}{t} \left[(1-\alpha) \int_0^t \frac{f(s)}{(t-s)^\alpha} ds - f(0) t^{1-\alpha} \right]$$

$$= \frac{1}{t} \int_0^t \frac{1}{(t-s)^\alpha} \cdot (t-s) f'(s) ds$$

$$= \frac{1}{t} \cdot \lim_{\varepsilon \searrow 0} \int_\varepsilon^t (t-s)^{1-\alpha} f'(s) ds. \tag{1.22}$$

This identity is the obvious consequence of applying the integration by parts for the integral in (1.22).

Let's return to (1.19) once more. To establish its validity, following [Samko et al. (1993)], start by recalling (1.14). We shall refer to the function there as $g = g(s,\tau)$.

Given a, $b > 0$, since g is locally absolutely integrable in the plane when f is merely absolutely continuous, Fubini's theorem yields — take $\gamma = 0$ in the formula of g —

$$\left(\int_0^t ds \int_0^s d\tau\right) \frac{f'(\tau)}{(t-s)^\alpha} = \int_0^t \left\{ \int_0^t [g(s,\tau)\chi_{(0,+\infty)}(s-\tau)]\, d\tau \right\} ds$$

$$= \int_{[0,t]^2} [g(s,\tau)\chi_{(0,+\infty)}(s-\tau)]\, ds d\tau$$

$$= \int_0^t \left\{ \int_0^t [g(s,\tau)\chi_{(0,+\infty)}(s-\tau)]\, ds \right\} d\tau$$

$$= \left(\int_0^t d\tau \int_\tau^t ds\right) \frac{f'(\tau)}{(t-s)^\alpha},$$

[Samko et al. (1993), p. 9]. Both parts of the identity can be given the simple representations as below,

$$\left(\int_0^t ds \int_0^s d\tau\right) \frac{f'(\tau)}{(t-s)^\alpha} = \int_0^t \left(\int_0^s f'(\tau)d\tau\right) \frac{ds}{(t-s)^\alpha} \qquad (1.23)$$

$$= \int_0^t \frac{f(s) - f(0)}{(t-s)^\alpha} ds$$

$$= -f(0) \cdot \frac{t^{1-\alpha}}{1-\alpha} + \int_0^t \frac{f(s)}{(t-s)^\alpha} ds \qquad (1.24)$$

and

$$\left(\int_0^t d\tau \int_\tau^t ds\right) \frac{f'(\tau)}{(t-s)^\alpha} = \int_0^t \left(\int_\tau^t \frac{ds}{(t-s)^\alpha}\right) f'(\tau)d\tau$$

$$= \frac{1}{1-\alpha} \int_0^t (t-\tau)^{1-\alpha} f'(\tau)d\tau. \qquad (1.25)$$

Notice that by putting together the right-hand parts of (1.24), (1.25), we have obtained the generalization of (1.22) to absolutely continuous functions.

We know already that the mapping $(s,\tau) \mapsto \frac{f'(\tau)}{(s-\tau)^\alpha} \cdot \chi_{(0,+\infty)}(s-\tau)$ is in $L^1_{\text{loc}}([0,+\infty)^2, \mathbb{R})$ — simply put $\alpha = 0$ and $\gamma = \alpha$ in the formula (1.14) —. Thus,

$$\left(\int_0^t ds \int_0^s d\tau\right) \frac{f'(\tau)}{(s-\tau)^\alpha} = \left(\int_0^t d\tau \int_\tau^t ds\right) \frac{f'(\tau)}{(s-\tau)^\alpha}. \qquad (1.26)$$

The right-hand member of (1.26) can be recast as

$$\left(\int_0^t d\tau \int_\tau^t ds\right) \frac{f'(\tau)}{(s-\tau)^\alpha} = \int_0^t f'(\tau) \cdot \frac{(t-\tau)^{1-\alpha}}{1-\alpha} d\tau, \qquad (1.27)$$

and so it is equal to the quantity in (1.25). A conclusion of this is the equality between the quantity in (1.24) and the left-hand member of (1.26). It reads as

$$-f(0) \cdot \frac{t^{1-\alpha}}{1-\alpha} + \int_0^t \frac{f(s)}{(t-s)^\alpha} ds = \left(\int_0^t ds \int_0^s d\tau \right) \frac{f'(\tau)}{(s-\tau)^\alpha} \quad (1.28)$$

$$= \int_0^t \left(\int_0^s \frac{f'(\tau)}{(s-\tau)^\alpha} d\tau \right) ds. \quad (1.29)$$

According to Fubini's theorem, the inner integral in (1.29) is a.e. in $L^1_{\text{loc}}([0, +\infty), \mathbb{R})$. Let's denote it with g and refer to its integral over $[0, t]$ by $G(t)$. We have

$$\left| \frac{G(t+h) - G(t)}{h} - g(t) \right| \le \frac{1}{h} \int_t^{t+h} |g(s) - g(t)| ds, \quad h > 0.$$

Observe that the right-hand part of the previous inequality occurs in the definition of the Lebesgue points of g, see [Rudin (1987), Chap. 7, Sect. 7.6]. Given the local integrability of g, the limit when $h \searrow 0$ of this quantity exists a.e. in $(0, +\infty)$ and is equal to 0. We conclude that *the function G is a.e. differentiable, its derivative is (a.e.) g and also $G(t + h) - G(t) = \int_t^{t+h} g(s)ds$ everywhere*. According to [Rudin (1987), Chap. 7, Th. 7.18 (c)], if we could add monotonicity of G to these features, then we could establish the absolute continuity of G. However, being an antiderivative of some L^1–function, that is g, *the function G is, in fact, absolutely continuous*, see [Kolmogorov and Fomine (1974), p. 339, Th. 2], as a consequence of the absolute continuity of the Lebesgue integral [Rudin (1987), Chap. 1, Ex. 12].

Thus, the left-hand member of (1.28) is (locally) absolutely continuous in $[0, +\infty)$ and its derivative is the inner integral from (1.29). So, differentiating both sides of the previous equality yields (1.19).

Let's make another claim in connection with the inquiry from page 6. Keeping the notations from that page, *if g is (locally) absolutely continuous then $\mathfrak{g}_\alpha$ is also (locally) absolutely continuous*. The proof of this assertion relies on (1.28), (1.29). In fact, since $t^{1-\alpha} = (1 - \alpha) \cdot \int_0^t \frac{ds}{s^\alpha}$, we deduce that the mapping $t \mapsto \int_0^t \frac{f(s)}{(t-s)^\alpha} ds$ is the difference of two antiderivatives of L^1_{loc}–functions, that is absolutely continuous.

We are at this point in possession of the fundamental identities for manipulating the Riemann-Liouville derivatives (1.5). Following [Abdeljawad et al. (2010)], several complementary estimates will be presented next. They can be established without significant input from measure theory.

Set β, $\gamma \in [0,1)$ and introduce the integral quantity below

$$(Q_{\beta,A}f)(t) = \int_a^t \frac{A(s)f(s)}{(t-s)^\beta}ds, \quad t > a, \tag{1.30}$$

where the functions $A : [a, +\infty) \to \mathbb{R}$ and $f : (a, +\infty) \to \mathbb{R}$ are continuous and f verifies the restriction

$$\lim_{t \searrow a}(t-a)^\gamma f(t) = f_a \in \mathbb{R}. \tag{1.31}$$

If $a = 0$ then $f \in \mathcal{RL}^\gamma$. Given that the Lebesgue measure is translation-invariant [Rudin (1987), Chap. 2, Th. 2.20], all of the previous computations remain valid.

We shall use the shorthand notations

$$\begin{cases} \|f\|_{L^\infty(a,\gamma;b,c)} = \max_{t \in [b,c]}[(t-a)^\gamma|f(t)|], \\ \|f\|_{L^\infty(a,\gamma;c)} = \sup_{t \in (a,c]}[(t-a)^\gamma|f(t)|] \end{cases} \tag{1.32}$$

for all $c > b > a$.

Obviously, $\|f\|_{L^\infty(a,\gamma;b,c)} \leq \|f\|_{L^\infty(a,\gamma;c)}$. It is easy to see that

$$|f_a| \leq \sup_{b \in (a,c)} \|f\|_{L^\infty(a,\gamma;b,c)} = \|f\|_{L^\infty(a,\gamma;c)}. \tag{1.33}$$

Notice[10] as well that — for $a < t \leq c$ and $\beta \in (0,1)$ —

$$\int_a^t \frac{|A(s)f(s)|}{(t-s)^\beta}ds = \int_a^t \frac{|A(s)| \cdot (s-a)^\gamma|f(s)|}{(t-s)^\beta(s-a)^\gamma}ds$$

$$\leq \int_a^t \frac{\|A\|_{L^\infty([a,c],\mathbb{R})}\|f\|_{L^\infty(a,\gamma;c)}}{(t-s)^\beta(s-a)^\gamma}ds$$

$$= (t-a)^{1-\beta-\gamma}B(1-\gamma,1-\beta)\|A\|_{L^\infty([a,c],\mathbb{R})}\|f\|_{L^\infty(a,\gamma;c)}. \tag{1.34}$$

We have employed the change of variables $s = a + \lambda(t-a)$ for $\lambda \in [0,1]$ and the Beta function.

Now, consider $a < b \leq t_1 < t_2 \leq c$. We get — again, via the change of variables $s = a + \lambda(t_2 - a)$ for $\lambda \in [0,1]$ —

$$\int_{t_1}^{t_2} \frac{ds}{(t_2-s)^\beta(s-a)^\gamma} = (t_2-a)^{1-\beta-\gamma} \cdot \int_{\frac{t_1-a}{t_2-a}}^1 \frac{d\lambda}{(1-\lambda)^\beta\lambda^\gamma} \tag{1.35}$$

$$= (t_2-a)^{1-\beta-\gamma} \cdot \int_0^1 \frac{1}{(1-\lambda)^\beta\lambda^\gamma} \cdot \chi_{\left[\frac{t_1-a}{t_2-a},1\right]}(\lambda)d\lambda$$

$$= \left[\int_0^{\frac{1}{2}} \frac{1}{(1-\lambda)^\beta\lambda^\gamma}\chi_{\left[\frac{t_1-a}{t_2-a},1\right]}(\lambda)d\lambda + \int_{\frac{1}{2}}^1 \frac{1}{(1-\lambda)^\beta\lambda^\gamma}\chi_{\left[\frac{t_1-a}{t_2-a},1\right]}(\lambda)d\lambda\right]$$

---

[10]Recall also (1.6)!

$$\times (t_2 - a)^{1-\beta-\gamma}$$
$$= (t_2 - a)^{1-\beta-\gamma} \cdot (I_1 + I_2). \tag{1.36}$$

For simplicity, write χ instead of χ_S from now on.

Introduce $p, q, v, w > 1$ such that $\frac{1}{p} + \frac{1}{q} = \frac{1}{v} + \frac{1}{w} = 1$ and $\gamma p, \beta v < 1$. An estimation of the integrals I_1, I_2 can be given by means of the Hölder inequality :

$$I_1 \leq \int_0^{\frac{1}{2}} \frac{1}{\left(\frac{1}{2}\right)^\beta \lambda^\gamma} \chi(\lambda) d\lambda = 2^\beta \int_0^{\frac{1}{2}} \left(\frac{1}{\lambda}\right)^\gamma \chi(\lambda) d\lambda$$

$$\leq 2^\beta \left[\int_0^{\frac{1}{2}} \left(\frac{1}{\lambda}\right)^{\gamma p} d\lambda \right]^{\frac{1}{p}} \cdot \left\{ \int_0^{\frac{1}{2}} [\chi(\lambda)]^q d\lambda \right\}^{\frac{1}{q}}$$

$$= 2^\beta \cdot \frac{1}{(1-\gamma p)^{\frac{1}{p}} \cdot 2^{\frac{1}{p}-\gamma}} \cdot \left\{ \int_0^{\frac{1}{2}} [\chi(\lambda)] d\lambda \right\}^{\frac{1}{q}}$$

$$= \frac{2^{\beta+\gamma-\frac{1}{p}}}{(1-\gamma p)^{\frac{1}{p}}} \cdot \left\{ \int_0^{\frac{1}{2}} [\chi(\lambda)] d\lambda \right\}^{\frac{1}{q}} \leq \frac{2^{\beta+\gamma-\frac{1}{p}}}{(1-\gamma p)^{\frac{1}{p}}} \cdot \left\{ \int_0^1 [\chi(\lambda)] d\lambda \right\}^{\frac{1}{q}}$$

$$= \frac{2^{\beta+\gamma-\frac{1}{p}}}{(1-\gamma p)^{\frac{1}{p}}} \cdot \left(1 - \frac{t_1-a}{t_2-a}\right)^{\frac{1}{q}} = c(p,\beta,\gamma) \cdot \left(\frac{t_2-t_1}{t_2-a}\right)^{\frac{1}{q}} \tag{1.37}$$

and

$$I_2 \leq 2^\gamma \int_{\frac{1}{2}}^1 \left(\frac{1}{1-\lambda}\right)^\beta \chi(\lambda) d\lambda$$

$$\leq 2^\gamma \left[\int_{\frac{1}{2}}^1 \frac{d\lambda}{(1-\lambda)^{\beta v}} \right]^{\frac{1}{v}} \cdot \left\{ \int_0^1 [\chi(\lambda)]^w d\lambda \right\}^{\frac{1}{w}}$$

$$= \frac{2^\gamma}{(1-\beta v)^{\frac{1}{v}} \cdot 2^{\frac{1}{v}-\beta}} \cdot \left(\frac{t_2-t_1}{t_2-a}\right)^{\frac{1}{w}}$$

$$= c(v,\gamma,\beta) \cdot \left(\frac{t_2-t_1}{t_2-a}\right)^{\frac{1}{w}}. \tag{1.38}$$

By taking into account (1.37), (1.38) and (1.36), we deduce that

$$\int_{t_1}^{t_2} \frac{ds}{(t_2-s)^\beta (s-a)^\gamma} \leq (t_2-a)^{1-\beta-\gamma}[c(p,\beta,\gamma) + c(v,\gamma,\beta)]$$

$$\times \max \left\{ \left(\frac{t_2-t_1}{t_2-a}\right)^{\frac{1}{q}}, \left(\frac{t_2-t_1}{t_2-a}\right)^{\frac{1}{w}} \right\}$$

$$= (t_2 - a)^{1-\beta-\gamma}[c(p,\beta,\gamma) + c(v,\gamma,\beta)] \cdot \left(\frac{t_2 - t_1}{t_2 - a}\right)^{\frac{1}{\max\{q,w\}}}$$

$$= \frac{C}{2}(t_2 - a)^{1-\beta-\gamma-\frac{1}{\max\{q,w\}}} \cdot (t_2 - t_1)^{\frac{1}{\max\{q,w\}}}, \tag{1.39}$$

where $C = C(p,v,\beta,\gamma) = 2[c(p,\beta,\gamma) + c(v,\gamma,\beta)]$.

Further, we would like to estimate the quantity

$$\int_a^{t_1} \frac{1}{(s-a)^\gamma} \cdot \left[\frac{1}{(t_1-s)^\beta} - \frac{1}{(t_2-s)^\beta}\right] ds,$$

where $\beta, \gamma \in (0,1)$ and $\beta + \gamma \leq 1$. The latter restriction will be used only at the very end of our estimation. Recall the notation $(\beta + \gamma - 1)^+ = \max\{\beta + \gamma - 1, 0\}$.

Before going further, remark that putting $A = t_1 - a$ and $B = t_2 - t_1$ in (1.8) leads to

$$0 \leq (t_2 - a)^\beta - (t_1 - a)^\beta \leq (t_2 - t_1)^\beta,$$

an estimate that will be used in the next series of computations.

Noticing that

$$B(1 - \gamma, 1 - \beta) = \frac{1}{(t_1 - a)^{1-\beta-\gamma}} \cdot \int_a^{t_1} \frac{ds}{(s-a)^\gamma(t_1-s)^\beta}$$

$$= \frac{1}{(t_2 - a)^{1-\beta-\gamma}} \cdot \int_a^{t_2} \frac{ds}{(s-a)^\gamma(t_2-s)^\beta},$$

we get

$$\int_a^{t_1} \frac{1}{(s-a)^\gamma} \cdot \left[\frac{1}{(t_1-s)^\beta} - \frac{1}{(t_2-s)^\beta}\right] ds$$

$$= \int_a^{t_1} \frac{ds}{(s-a)^\gamma(t_1-s)^\beta} - \int_a^{t_2} \frac{ds}{(s-a)^\gamma(t_2-s)^\beta}$$

$$+ \int_{t_1}^{t_2} \frac{ds}{(t_2-s)^\beta(s-a)^\gamma} = I_3 + I_4.$$

The integral I_4 was already evaluated, see (1.39).

We have — remember that $t_1 < t_2$ —

$$I_3 = B(1 - \gamma, 1 - \beta) \cdot [(t_1 - a)^{1-\beta-\gamma} - (t_2 - a)^{1-\beta-\gamma}] \tag{1.40}$$

$$= B(1 - \gamma, 1 - \beta) \cdot \frac{(t_2 - a)^{\beta+\gamma-1} - (t_1 - a)^{\beta+\gamma-1}}{[(t_2 - a)(t_1 - a)]^{\beta+\gamma-1}}$$

$$\leq \text{sign}\left[(\beta + \gamma - 1)^+\right] \cdot \left[\frac{(t_2 - t_1)}{(t_1 - a)^2}\right]^{\beta+\gamma-1}$$

$$= \delta(t_1 - a)^{2(1-\beta-\gamma)} \cdot (t_2 - t_1)^{\beta+\gamma-1}, \quad \delta \in \{0, 1\}.$$

So, via the idempotency of δ, if $\beta + \gamma \le 1$,

$$\int_a^{t_1} \frac{1}{(s-a)^\gamma} \cdot \left[\frac{1}{(t_1-s)^\beta} - \frac{1}{(t_2-s)^\beta} \right] ds = I_3 + I_4$$

$$\le \left[\frac{C}{2}(t_2-a)^{1-\beta-\gamma-\frac{1}{\max\{q,w\}}} + \delta \cdot (t_1-a)^{2(1-\beta-\gamma)} \right]$$

$$\times \max \left\{ (t_2-t_1)^{\frac{1}{\max\{q,w\}}}, \delta \cdot (t_2-t_1)^{\beta+\gamma-1} \right\} \tag{1.41}$$

$$\le \frac{C}{2}(t_2-a)^{1-\beta-\gamma-\frac{1}{\max\{q,w\}}} \cdot (t_2-t_1)^{\frac{1}{\max\{q,w\}}}. \tag{1.42}$$

Finally, take $\gamma, \beta \in (0, 1)$ such that $\beta + \gamma \le 1$, and $f \in C((a, +\infty), \mathbb{R})$ satisfying (1.31), and $t_1 < t_2$ in $[b, c]$ for $b > a$, and observe that

$$|(Q_{\beta,A}f)(t_1) - (Q_{\beta,A}f)(t_2)|$$

$$= \left| \int_a^{t_1} [A(s) \cdot (s-a)^\gamma f(s)] \cdot \frac{1}{(s-a)^\gamma} \left[\frac{1}{(t_1-s)^\beta} - \frac{1}{(t_2-s)^\beta} \right] ds \right.$$

$$\left. - \int_{t_1}^{t_2} [A(s) \cdot (s-a)^\gamma f(s)] \cdot \frac{ds}{(t_2-s)^\beta(s-a)^\gamma} \right|$$

$$\le \|A\|_{L^\infty([a,c],\mathbb{R})} \|f\|_{L^\infty(a,\gamma;b,c)} \cdot \int_{t_1}^{t_2} \frac{ds}{(t_2-s)^\beta(s-a)^\gamma} \tag{1.43}$$

$$+ \|A\|_{L^\infty([a,c],\mathbb{R})} \|f\|_{L^\infty(a,\gamma;c)} \int_a^{t_1} \frac{1}{(s-a)^\gamma} \cdot \left[\frac{1}{(t_1-s)^\beta} - \frac{1}{(t_2-s)^\beta} \right] ds$$

$$\le \left[C(t_2-a)^{1-\beta-\gamma-\frac{1}{\max\{q,w\}}} + \delta \cdot (t_1-a)^{2(1-\beta-\gamma)} \right]$$

$$\times \max \left\{ (t_2-t_1)^{\frac{1}{\max\{q,w\}}}, \delta \cdot (t_2-t_1)^{\beta+\gamma-1} \right\} \tag{1.44}$$

$$\times \|A\|_{L^\infty([a,c],\mathbb{R})} \|f\|_{L^\infty(a,\gamma;c)}$$

$$= C \cdot (t_2-a)^{1-\beta-\gamma-\frac{1}{\max\{q,w\}}}$$

$$\times (t_2-t_1)^{\frac{1}{\max\{q,w\}}} \cdot \|A\|_{L^\infty([a,c],\mathbb{R})} \|f\|_{L^\infty(a,\gamma;c)}, \tag{1.45}$$

by means of (1.39), (1.42).

Now, given $\xi \in [0, 1)$,

$$|(t_1-a)^\xi(Q_{\beta,A}f)(t_1) - (t_2-a)^\xi(Q_{\beta,A}f)(t_2)|$$

$$= \left| (t_1-a)^\xi[(Q_{\beta,A}f)(t_1) - (Q_{\beta,A}f)(t_2)] \right.$$

$$\left. - (Q_{\beta,A}f)(t_2)[(t_2-a)^\xi - (t_1-a)^\xi] \right|$$

$$\le (t_1-a)^\xi \cdot |(Q_{\beta,A}f)(t_1) - (Q_{\beta,A}f)(t_2)| \tag{1.46}$$

$$+ |(Q_{\beta,A}f)(t_2)| \cdot (t_2 - t_1)^{\xi}$$

$$\leq (t_1 - a)^{\xi} \cdot \left[C(t_2 - a)^{1-\beta-\gamma-\frac{1}{\max\{q,w\}}} + \delta \cdot (t_1 - a)^{2(1-\beta-\gamma)} \right]$$

$$\times \max \left\{ (t_2 - t_1)^{\frac{1}{\max\{q,w\}}}, \delta \cdot (t_2 - t_1)^{\beta+\gamma-1} \right\}$$

$$\times \|A\|_{L^{\infty}([a,c],\mathbb{R})} \|f\|_{L^{\infty}(a,\gamma;c)}$$

$$+ (t_2 - a)^{1-\beta-\gamma} B(1-\gamma, 1-\beta) \cdot \|A\|_{L^{\infty}([a,c],\mathbb{R})} \|f\|_{L^{\infty}(a,\gamma;c)}$$

$$\times (t_2 - t_1)^{\xi}$$

$$\leq (c-a)^{\xi} \left[C(t_2 - a)^{1-\beta-\gamma-\frac{1}{\max\{q,w\}}} + \delta \cdot (t_1 - a)^{2(1-\beta-\gamma)} \right. \qquad (1.47)$$

$$+ (t_2 - a)^{1-\beta-\gamma} \cdot B(1-\gamma, 1-\beta) \right] \|A\|_{L^{\infty}([a,c],\mathbb{R})} \|f\|_{L^{\infty}(a,\gamma;c)} \qquad (1.48)$$

$$\times \max \left\{ (t_2 - t_1)^{\frac{1}{\max\{q,w\}}}, \delta \cdot (t_2 - t_1)^{\beta+\gamma-1}, (t_2 - t_1)^{\xi} \right\} \qquad (1.49)$$

$$\leq D \|A\|_{L^{\infty}([a,c],\mathbb{R})} \|f\|_{L^{\infty}(a,\gamma;c)}$$

$$\times \max \left\{ (t_2 - t_1)^{\frac{1}{\max\{q,w\}}}, (t_2 - t_1)^{\xi} \right\}, \qquad (1.50)$$

where $D = D(p, v, \beta, \gamma, b-a, c-a)$. We took into account the estimates (1.34), (1.44).

Let us make a comment regarding D and (1.47), (1.48). If we don't know the signs of the quantities $1 - \beta - \gamma - \frac{1}{\max\{q,w\}}$ and $1 - \beta - \gamma$ then we are forced to use the raw estimates

$$(t_2 - a)^{1-\beta-\gamma-\frac{1}{\max\{q,w\}}} \leq (b-a)^{1-\beta-\gamma-\frac{1}{\max\{q,w\}}} + (c-a)^{1-\beta-\gamma-\frac{1}{\max\{q,w\}}}$$

and its siblings[11] for $(t_2 - a)^{1-\beta-\gamma}$ and $(t_1 - a)^{2(1-\beta-\gamma)}$, which will reflect on the size of D. If, on the other hand, we have $1 - \beta - \gamma - \frac{1}{\max\{q,w\}} > 0$ — which implies that $1 - \beta - \gamma > 0$ — then

$$(t_2 - a)^{1-\beta-\gamma-\frac{1}{\max\{q,w\}}} \leq (c-a)^{1-\beta-\gamma-\frac{1}{\max\{q,w\}}}, \quad t_2 \in [b, c],$$

and so

$$D = C \cdot (c-a)^{1+\xi-\beta-\gamma-\frac{1}{\max\{q,w\}}} + (c-a)^{1-\beta-\gamma} \cdot B(1-\gamma, 1-\beta). \qquad (1.51)$$

Observe that D is now *independent* of b, a fact of crucial importance for our investigation of a fractional Fite theorem [Abdeljawad et al. (2010), Theorem 3.1].

In conclusion, according to (1.41), (1.44), *the function* $Q_{\beta,A}f \in C((a, +\infty), \mathbb{R})$ *is locally Hölder continuous whenever* f *is subjected to (1.31) for some* $\beta, \gamma \in (0, 1)$. The same thing is valid, via (1.49), for the mapping $t \mapsto (t-a)^{\xi}(Q_{\beta,A})f(t)$ when $\xi \in (0, 1)$.

---

[11]The quantity $\delta(t_1 - a)^{2(1-\beta-\gamma)}$ is majorized by $\left(\frac{1}{b-a} \right)^{2(\beta+\gamma-1)}$ when $\beta + \gamma > 1$.

For the sake of completeness, let us state some important details about the case when $A \equiv 1$. First, *one has* $\lim_{\beta \searrow 1}(Q_{\beta,1}f)(t) = f(t)$ *for all the Lebesgue points of f* [Samko et al. (1993), p. 51, Theorem 2.7], *or everywhere if $f \in C([0, +\infty), \mathbb{R})$* [Podlubny (1999a), p. 66, 67]. Second, and this is a Hardy-Littlewood theorem [Samko et al. (1993), p. 66, Theorem 3.5, and pp. 91, 92], *if $f \in L^p((0, T), \mathbb{R})$ for some $p \in \left(1, \frac{1}{\beta}\right)$ then $Q_{\beta,1}f \in L^q((0, T), \mathbb{R})$, where $q = \frac{p}{1 - \beta p}$ and $T > 0$.*

1.3 The Abel Computation

Set $\zeta \in [0, 1)$, $\alpha \in (0, 1)$ and $g \in C((a, b), \mathbb{R})$ which satisfies (1.31) for $\gamma = \zeta$. We are interested in finding some function $f \in C((a, b), \mathbb{R})$ subjected to the restriction (1.31)[12] for $\gamma = 1 - \alpha$ such that

$$((_aD_t^\alpha f)(t) =) \frac{1}{\Gamma(1 - \alpha)} \cdot \frac{d}{dt}\left[\int_a^t \frac{f(s)}{(t - s)^\alpha}ds\right] = g(t), \quad t \in (a, b). \quad (1.52)$$

Let us return to (1.45), where $\beta = 0$ and $A \equiv 1$. We may take $w > 1$ so large that $1 - \beta - \gamma - \frac{1}{\max\{q, w\}} = 1 - \zeta - \frac{1}{w} > 0$. Also, D in (1.51) — for $\xi = 0$ — is independent of b, and we deduce that *the function $Q_{0,1}g$ is locally Hölder continuous in $[a, b)$ and, via (1.34), $(Q_{0,1}g)(a) = 0$.*

Moreover, being the antiderivative of an L_{loc}^1–function, $Q_{0,1}g$ is *locally absolutely continuous in $[a, b)$*. Its derivative is, obviously, the function g since

$$(Q_{0,1}g)(t) = \left(\int_a^{a+\varepsilon} + \int_{a+\varepsilon}^t\right)g(s)ds = c_\varepsilon + g_\varepsilon(t), \quad c_\varepsilon \in \mathbb{R}, \quad (1.53)$$

and $g_\varepsilon \in C^1([a + \varepsilon, t], \mathbb{R})$ with $g_\varepsilon' = g$, see [Kolmogorov and Fomine (1974), p. 301, Th. 9]. As $g \in \mathcal{RL}^\zeta$, it is evident that $\lim_{t \searrow a}[(t - a)^{1+\beta} \cdot g(t)] = \lim_{t \searrow a}[(t - a)^\zeta g(t) \cdot (t - a)^{(1-\zeta)+\beta}] = 0$ for all $\beta \geq 0$ and this yields $(Q_{0,1}g)' \in \mathcal{RL}^{1+\beta}$. Recalling the restrictions needed for the existence of (1.11), we conclude as well that *the Riemann-Liouville derivative of $Q_{0,1}g$, of any order $\alpha \in (0, 1)$, exists everywhere in (a, b).* If, on the other hand, the function g would be only integrable then $Q_{0,1}g$ would be merely absolutely continuous (locally) and *its Riemann-Liouville derivative of order α would exist only a.e.* by means of (1.19).

---

[12] Allow us to refer to this property in a loose manner by saying that $f \in \mathcal{RL}^\gamma$ when there is no danger of confusion regarding the existence interval of f.

The absolute continuity of $Q_{0,1}g$ implies the absolute continuity of $Q_{1-\alpha,1}(Q_{0,1}g)$, see page 13. Another application of (1.34) leads to $\lim\limits_{t \searrow a} Q_{1-\alpha,1}(Q_{0,1}g)(t) = 0$ and so, following [Samko et al. (1993), p. 31, Th. 2.1], *the Abel integral equation with the free term $Q_{0,1}g$ is solvable in $L^1((a,b), \mathbb{R})$.*

To solve (1.52), we recast it first as

$$\frac{1}{\Gamma(1-\alpha)} \left\{ \int_a^t \frac{f(s)}{(t-s)^\alpha} ds - \lim_{t \searrow a} \left[\int_a^t \frac{f(s)}{(t-s)^\alpha} ds \right] \right\} = (Q_{0,1}g)(t). \quad (1.54)$$

By taking into account the simple estimate[13]

$$\left| \int_a^t \frac{1}{(t-s)^\alpha (s-a)^{1-\alpha}} \cdot [(s-a)^{1-\alpha} f(s) - f_a] ds \right|$$
$$\leq B(\alpha, 1-\alpha) \cdot \sup_{s \in (a,t]} |(s-a)^{1-\alpha} f(s) - f_a|, \quad (1.55)$$

the equality (1.54) yields

$$\int_a^t \frac{f(s)}{(t-s)^\alpha} ds - f_a \cdot B(\alpha, 1-\alpha) = \Gamma(1-\alpha) \cdot (Q_{0,1}g)(t).$$

Further, we have

$$\int_a^x \frac{1}{(x-t)^{1-\alpha}} \int_a^t \frac{f(s)}{(t-s)^\alpha} ds\, dt - f_a B(\alpha, 1-\alpha) \cdot \frac{(x-a)^\alpha}{\alpha}$$
$$= \Gamma(1-\alpha) \int_a^x \frac{1}{(x-t)^{1-\alpha}} \int_a^t g(s) ds\, dt. \quad (1.56)$$

Next, following an argumentation similar to that leading to (1.26), (1.27), the left-hand part of the preceding relation becomes

$$\int_a^x f(s) \int_s^x \frac{dt}{(x-t)^{1-\alpha}(t-s)^\alpha} ds - \frac{f_a B(\alpha, 1-\alpha)}{\alpha} \cdot (x-a)^\alpha$$
$$= B(\alpha, 1-\alpha) \cdot \left[\int_a^x f(s) ds - \frac{f_a}{\alpha} \cdot (x-a)^\alpha \right], \quad (1.57)$$

while the right-hand member reads as

$$\Gamma(1-\alpha) \int_a^x \int_a^t \frac{g(s)}{(t-s)^{1-\alpha}} ds\, dt = \Gamma(1-\alpha) \int_a^x (Q_{1-\alpha,1}g)(t) dt$$

after repeating verbatim the proof in (1.23), (1.29).

---

[13]If $f \in \mathcal{RL}^\beta$ for some $\beta \in [0, 1-\alpha)$ then, obviously, $f_a = 0$.

By differentiating with respect to x, where $x > a$, the equality — observe that, following (1.3), we have $B(\alpha, 1 - \alpha) = B(1 - \alpha, \alpha) = \Gamma(\alpha) \cdot \Gamma(1 - \alpha)$ — below

$$\Gamma(\alpha) \cdot \left[\int_a^x f(s)ds - \frac{f_a(x-a)^\alpha}{\alpha} \right] = \int_a^x (Q_{1-\alpha,1}g)(t)dt,$$

we get the following *integral counterpart*[14] of the Riemann-Liouville derivative of f, namely[15]

$$f(x) = \frac{f_a}{(x-a)^{1-\alpha}} + \frac{1}{\Gamma(\alpha)} \int_a^x \frac{g(s)}{(x-s)^{1-\alpha}}ds, \quad x > a. \tag{1.58}$$

It remains to establish that $f \in \mathcal{RL}^{1-\alpha}$. First, by taking into account either of the estimates (1.44), (1.49) for $\xi = \beta = 1 - \alpha$ and $\gamma = \zeta$, we observe that *the right-hand part of (1.58) is locally Hölder continuous in* (a, b). Second, via (1.34), $\lim\limits_{x \searrow a} [(x-a)^{1-\alpha}(Q_{1-\alpha,1}g)(x)] = \lim\limits_{x \searrow a} (x-a)^{1-\zeta} \cdot B(1-\zeta, \alpha) \|g\|_{L^\infty(a,\zeta;c)} = 0$, and so f satisfies (1.31).

This concludes our *Abel computation* regarding the solvability of (1.52) since it finds a convenient f, given by (1.58), that verifies (1.52) in (a, b).

1.4   The Operators. The Caputo Differential

The three differential operators of order $1 + \alpha$ we shall use in the asymptotic analysis are built on the "skeleton" (1.5). They can be stated as

$$_0^1\mathcal{O}_t^{1+\alpha} = {_0}D_t^\alpha \circ \frac{d}{dt}, \quad _0^2\mathcal{O}_t^{1+\alpha} = \frac{d}{dt} \circ {_0}D_t^\alpha,$$

and

$$_0^3\mathcal{O}_t^{1+\alpha} = {_0}D_t^\alpha \circ \left(t \cdot \frac{d}{dt} - \mathrm{id}_{\mathcal{RL}^\alpha} \right),$$

see [Baleanu et al. (2011b)].

The operator $_0^3\mathcal{O}_t^{1+\alpha}$ is based upon the following splitting of the second order differential operator $\frac{d^2}{dt^2}$, namely

$$tx'' = (tx' - x)' = [tx' - x + x(0)]', \quad t > 0,$$

---

[14]This computation leading to (1.58) reduces, from the viewpoint of functional analysis, to finding the formula of $f = T^{-1}(g)$ when $T(f) = g$. As T is a *differential* operator, T^{-1} will be an *integral* operator. We prefer, however, the friendlier approach given by either of the expressions *integral counterpart* and *integral representation* of T when referring to T^{-1}.

[15]Normally, the identity (1.58) should be valid a.e.. However, since both parts of the identity consist of continuous functions, if they agree a.e. on (a, b) then they agree everywhere. A proof of this assertion can be found at page 38.

which stems from the integration technique in the Lie algebra L_2, see [Ibragimov and Kovalev (2009), p. 23], [Baleanu et al. (2010c), p. 3]. On the other hand, the operators ${}_0^1\mathcal{O}_t^{1+\alpha}$ and ${}_0^2\mathcal{O}_t^{1+\alpha}$ give *fractionalizations* of the natural factorization of operator $\frac{d^2}{dt^2}$, that is $x'' = (x')'$.

The quantity

$$
{}_0C_t^\alpha(f)(t) = \left(\text{Int} \circ {}_0^1\mathcal{O}_t^{1+\alpha}\right)(f)(t)
$$

$$
= \frac{1}{\Gamma(1-\alpha)} \cdot \int_0^t \frac{f'(s)}{(t-s)^\alpha}ds, \quad t > 0, \tag{1.59}
$$

where $f \in C([0,+\infty),\mathbb{R})$ is absolutely continuous and[16] — recall (1.6) — $s^\beta f' \in L_{\text{loc}}^\infty([0,+\infty),\mathbb{R})$ for some $\beta \in [0,1)$, often bears the name *Caputo differential operator* [17], of *order* $\alpha \in (0,1)$ [Podlubny (1999a), p. 78]. An obvious particular case is when $f \in C^1((0,+\infty),\mathbb{R}) \cap C([0,+\infty),\mathbb{R})$ and $f' \in \mathcal{RL}^\beta$, where $\beta \in [0,1)$.

To produce an integral representation for the Caputo differential, let us return to the circumstances of (1.54). Assuming that g is (locally) absolutely continuous, this equality is adapted in our case as

$$
\int_a^t \frac{f'(s)}{(t-s)^\alpha}ds = \Gamma(1-\alpha) \cdot g(t), \quad t > a.
$$

Also, the identities (1.56), (1.57) lead to

$$
B(\alpha, 1-\alpha) \int_a^x f'(s)ds = \Gamma(1-\alpha) \int_a^x \frac{g(t)}{(x-t)^{1-\alpha}}dt.
$$

Finally, the formula below is the homologous of (1.58)

$$
f(x) = f(a) + \frac{1}{\Gamma(\alpha)} \int_a^x \frac{g(s)}{(x-s)^{1-\alpha}}ds, \quad x > a. \tag{1.60}
$$

The Caputo differential operator is sometimes preferred to the Riemann-Liouville derivative in the modeling of physical phenomena because the *initial value* of f from the preceding formula can be given a natural explanation, which is not the case for the data in (1.58) [Podlubny (1999a), ibid.].

In various situations, the Caputo differential operator can be *approximated*[18] by a modified version of (1.5), namely

$$
{}_a\mathbb{D}_t^\alpha(f)(t) = \frac{1}{\Gamma(1-\alpha)} \cdot \frac{d}{dt}\left[\int_a^t \frac{f(s)-f(a)}{(t-s)^\alpha}ds\right], \quad t > 0,
$$

---

[16]As already noted, the restriction regarding f' is not necessary though it simplifies computations.

[17]Or, *Caputo differential* for short.

[18]By this we mean having the same integral representation or integral counterpart.

see [Baleanu and Mustafa (2009), p. 2], where $f \in C([a, +\infty), \mathbb{R})$ is absolutely continuous and $s^{1+\beta} f' \in L_{\text{loc}}^{\infty}([a, +\infty), \mathbb{R})$ for some $\beta \geq 0$. Since $f - f(a) \in \mathcal{RL}^0$ and $[f - f(a)]_a = 0$ in (1.31) for $\gamma = 0$, the integral counterpart (1.58) of the Riemann-Liouville derivative of $f - f(a)$ reduces to the integral counterpart (1.60) of Caputo's operator.

1.5   The Integral Representation of the Operators. The Half-line Case

Following [Baleanu et al. (2010b, 2011a,b)], we shall construct the integral formulas — similar to (1.58), (1.60) — which represent the operators ${}_{0}^{i}\mathcal{O}_{t}^{1+\alpha}$, where $i \in \{1, 2, 3\}$, for functions defined on $[0, +\infty)$.

Take $h \in C([0, +\infty), \mathbb{R}) \cap L^1((0, +\infty), \mathbb{R})$. The identity

$${}_{0}^{1}\mathcal{O}_{t}^{1+\alpha}(x)(t) = h(t), \quad t > 0,$$

is recast as

$$\frac{1}{\Gamma(1 - \alpha)} \int_{0}^{t} \frac{x'(s)}{(t - s)^{\alpha}} ds = x_1 - \int_{t}^{+\infty} h(s) ds$$
$$= g(t), \quad t > 0,$$

where $x_1 = \lim_{t \to +\infty} {}_{0}C_{t}^{\alpha}(x)(t) \in \mathbb{R}$.

Since g verifies (1.31) for $\gamma = 0$ and $g_0 = g(0)$, we may use the formula (1.60) to obtain the *integral representation of* ${}_{0}^{1}\mathcal{O}_{t}^{1+\alpha}$, that is

$$x(t) = x_0 + x_2 \cdot t^{\alpha} - \frac{1}{\Gamma(\alpha)} \int_{0}^{t} \frac{1}{(t - s)^{1-\alpha}} \int_{s}^{+\infty} h(\tau) d\tau ds, \quad (1.61)$$

where $x_0 = x(0)$ and $x_2 = \frac{x_1}{\Gamma(1+\alpha)}$. We have employed the estimate $\alpha \cdot \Gamma(\alpha) = \Gamma(1 + \alpha)$, recall (1.3).

Now, consider the equation

$${}_{0}^{2}\mathcal{O}_{t}^{1+\alpha}(x)(t) = h(t), \quad t > 0.$$

Its first integro-differential reformulation reads as

$${}_{0}D_{t}^{\alpha}(x)(t) = x_1 - \int_{t}^{+\infty} h(s) ds, \quad t > 0,$$

where $x_1 = \lim_{t \to +\infty} {}_{0}D_{t}^{\alpha}(x)(t) \in \mathbb{R}$.

The integral counterpart (1.58) of the Riemann-Liouville derivative leads to

$$x(t) = x_0 \cdot t^{\alpha-1} + x_2 \cdot t^{\alpha} - \frac{1}{\Gamma(\alpha)} \int_{0}^{t} \frac{1}{(t - s)^{1-\alpha}} \int_{s}^{+\infty} h(\tau) d\tau ds, \quad (1.62)$$

where x_0 is given by (1.31) for $\gamma = 1 - \alpha$ and $x_2 = \frac{x_1}{\Gamma(1+\alpha)}$. To see that x_0 has been correctly defined, notice that

$$\left| \int_0^t \frac{1}{(t-s)^{1-\alpha}} \int_s^{+\infty} h(\tau) d\tau ds \right| \leq \frac{t^\alpha}{\alpha} \cdot \|h\|_{L^1((0,+\infty),\mathbb{R})}, \quad t > 0,$$

and so $\lim_{t \searrow 0} [t^{1-\alpha} \cdot x(t)] = x_0 + \lim_{t \searrow 0} \left\{ t \cdot \left[|x_2| + \frac{1}{\Gamma(1+\alpha)} \|h\|_{L^1((0,+\infty),\mathbb{R})} \right] \right\} = x_0.$

Formula (1.62) constitutes the *integral representation of* ${}_0^2\mathcal{O}_t^{1+\alpha}$.

Finally, to deal with the equation

$$ {}_0^3\mathcal{O}_t^{1+\alpha}(x)(t) = h(t), \quad t > 0,$$

introduce the relations

$$y(t) = tx'(t) - x(t), \quad x(t) = c \cdot t - t \int_t^{+\infty} \frac{y(\tau)}{\tau^2} d\tau, \quad t > 0, \quad (1.63)$$

with $c \neq 0$ and $y \in \mathcal{RL}^\gamma$ for some $\gamma \geq 0$, see [Baleanu et al. (2010c)].

As before,

$$\frac{1}{\Gamma(1-\alpha)} \int_0^t \frac{y(s)}{(t-s)^\alpha} ds = x_1 - \int_t^{+\infty} h(s) ds, \quad t > 0,$$

where $x_1 = \lim_{t \to +\infty} \frac{1}{\Gamma(1-\alpha)} \int_0^t \frac{y(s)}{(t-s)^\alpha} ds \in \mathbb{R}$, and, repeating the inferences drawn from (1.56), we obtain that

$$\int_0^t y(s) ds$$

$$= \frac{x_1 t^\alpha}{\Gamma(1+\alpha)} - \frac{1}{\Gamma(\alpha)} \int_0^t \frac{1}{(t-s)^{1-\alpha}} \int_s^{+\infty} h(\tau) d\tau ds$$

$$= \frac{x_1 t^\alpha}{\Gamma(1+\alpha)} - \frac{1}{\Gamma(\alpha)} \int_0^t \frac{1}{(t-s)^{1-\alpha}} \left(\int_0^{+\infty} - \int_0^s \right) h(\tau) d\tau ds$$

$$= \frac{t^\alpha}{\Gamma(1+\alpha)} \left[x_1 - \int_0^{+\infty} h(\tau) d\tau \right] + \frac{1}{\Gamma(\alpha)} \int_0^t \int_0^s \frac{h(u)}{(s-u)^{1-\alpha}} du ds.$$

We have used the identity

$$\int_0^t \frac{1}{(t-s)^{1-\alpha}} \int_0^s h(\tau) d\tau ds = \int_0^t \int_0^s \frac{h(u)}{(s-u)^{1-\alpha}} du ds,$$

which is a consequence of (1.23), (1.25), (1.27), (1.29) by taking $f' = h$.

By differentiation, the latter formula leads to

$$y(t) = \frac{t^{\alpha-1}}{\Gamma(\alpha)} \left[x_1 - \int_0^{+\infty} h(\tau) d\tau \right] + \frac{1}{\Gamma(\alpha)} \int_0^t \frac{h(s)}{(t-s)^{1-\alpha}} ds,$$

where $t > 0$.

In conclusion, the *integral representation of* ${}_0^3\mathcal{O}_t^{1+\alpha}$ is given by the set of equations

$$\begin{cases} x(t) = t \cdot \left[c_1 - \int_t^{+\infty} \frac{y(\tau)}{\tau^2} d\tau \right], & c_1 \in \mathbb{R}, \\ y(t) = \frac{c_2}{t^{1-\alpha}} + \frac{1}{\Gamma(\alpha)} \int_0^t \frac{h(s)}{(t-s)^{1-\alpha}} ds, \end{cases} \tag{1.64}$$

where $c_2 = \frac{1}{\Gamma(\alpha)} \left[x_1 - \int_0^{+\infty} h(\tau) d\tau \right]$.

Observe that, for the integral in the equation of x to be convergent, we must control the size of y in $[t, +\infty)$. By taking into account the estimate

$$\left| \int_0^t \frac{h(s)}{(t-s)^{1-\alpha}} ds \right| \leq \|h\|_{L^\infty([0,t],\mathbb{R})} \cdot \frac{t^\alpha}{\alpha}, \quad t > 0,$$

we shall ask from h to obey as well the inequality

$$\int_t^{+\infty} \frac{\|h\|_{L^\infty([0,s],\mathbb{R})}}{s^{2-\alpha}} ds < +\infty, \quad t > 0.$$

An obvious particular case is when $h \in C([0, +\infty), \mathbb{R}) \cap (L^1 \cap L^\infty)$ $((0, +\infty), \mathbb{R})$.

Chapter 2

Existence and Uniqueness of Solution for the Differential Equations of Order α

Let us introduce the initial value problem

$$\begin{cases} {_0}C_t^\alpha(x)(t) = f(t, x(t)), \, t > 0, \\ x(0) = x_0, \qquad\qquad x_0 \in \mathbb{R}, \end{cases} \tag{2.1}$$

where the function $f : \mathbb{R}_+ \times \mathbb{R} \to \mathbb{R}$ is assumed continuous. We shall employ sometimes $\mathbb{R}_+$ as a shorthand notation for the non-negative half-axis.

The Caputo differential operator ${_0}C_t^\alpha$ has been displayed in (1.59) and we recall its integral representation (1.60) which becomes now

$$x(t) = x_0 + \frac{1}{\Gamma(\alpha)} \int_0^t \frac{f(s, x(s))}{(t - s)^{1-\alpha}} ds, \quad t \geq 0. \tag{2.2}$$

The first result in this chapter deals with existence and uniqueness *in the large*, that is in $\mathbb{R}_+$.

Theorem 2.1. ([Baleanu and Mustafa (2010a), Theorem 2]) *Assume that there exists the continuous function* $F : \mathbb{R}_+ \to \mathbb{R}_+$ *such that*

$$|f(t, x) - f(t, y)| \leq F(t)|x - y|, \quad t \geq 0 \quad and \quad x, y \in \mathbb{R}.$$

Then, the integral equation (2.2) has a unique continuous solution defined everywhere in $\mathbb{R}_+$.

This theorem of uniqueness is a fundamental tool in the case of ordinary differential equations — put $\alpha = 1$ in (2.2) —. The book by Kartsatos [Kartsatos (1980)], similarly to many other monographs devoted to the *qualitative theory* of ordinary differential equations, contains a detailed presentation of the typical proof. The latter consists of introducing a complete metric space that will be endowed with an exponentially weighted metric (the so-called Bielecki, or Morgenstern-Bielecki [Walter (1998)], metric), followed by a verification of the claim that a certain integral operator acting on the metric space is a contraction. The metric space is given by families of continuous functions defined on a bounded interval of *arbitrary length*. This is the key feature which allows one to say that *a solution of (2.2) whose existence can be established on any compact subinterval of* $\mathbb{R}_+$ *exists, naturally, throughout* $\mathbb{R}_+$.

In his rapid demonstration of the Picard-Lindelöf theorem of existence and uniqueness of the solution to a Cauchy problem, Brezis [Brezis (1999)] shows how a *prospective study of the growth* of a solution to (2.2) will lead to establishing the existence of solutions throughout $\mathbb{R}_+$ *at once*, that is, without the unpleasant artifice of bringing into the study the arbitrarily long compact subintervals of $\mathbb{R}_+$. His proof, however, regarded a particular case of the nonlinearity $f(t, x)$. The trick needed in the general circumstances is more involved and it will be given in the following in the unifying context of the integral equation (2.2).

Proof. Define two continuous functions h, $H_\lambda : \mathbb{R}_+ \to \mathbb{R}_+$ with the formulas

$$h(t) = 1 + |x_0| + \frac{1}{\Gamma(\alpha)} \int_0^t \frac{|f(s, x_0)|}{(t - s)^{1-\alpha}} ds$$

and

$$H_\lambda(t) = h(t) \cdot \exp\left(t + \frac{\lambda}{q} \int_0^t [h(s)F(s)]^q ds\right)$$

for a fixed $\lambda > 0$. Here, q is taken such that $\frac{1}{p} + \frac{1}{q} = 1$ and $1 < p < \min\left\{\frac{1}{\alpha}, \frac{1}{1-\alpha}\right\}$.

Consider further the metric space $\mathcal{X} = (X, d_\lambda)$, where X is the set of all functions $x \in C(\mathbb{R}_+, \mathbb{R})$ having the asymptotic behavior given by $x(t) = O(H_\lambda(t))$ when $t \to +\infty$ and

$$d_\lambda(x, y) = \sup_{t \geq 0} \left\{ \frac{|x(t) - y(t)|}{H_\lambda(t)} \right\} \quad \text{for any } x, y \in X. \tag{2.3}$$

Since $H_\lambda(t) \geq h(t) \geq 1$, *the constant functions belong to X*. To establish that the metric space is complete, one may follow the presentations in [Corduneanu (1973), p. 2] or [Walter (1998), p. 61, 349].

Given the operator $T : \mathcal{X} \to C(\mathbb{R}_+, \mathbb{R})$ with the formula

$$(Tx)(t) = x_0 + \frac{1}{\Gamma(\alpha)} \int_0^t \frac{f(s, x(s))}{(t - s)^{1-\alpha}} ds,$$

we have the following estimates as a consequence of applying the Hölder inequality

$$|(Tx)(t) - (Ty)(t)| \leq \frac{1}{\Gamma(\alpha)} \int_0^t \frac{F(s)}{(t - s)^{1-\alpha}} |x(s) - y(s)| ds$$

$$= \frac{1}{\Gamma(a)} \int_0^t \frac{e^s}{(t-s)^{1-\alpha}} \cdot F(s) \frac{|x(s) - y(s)|}{e^s} ds$$

$$\leq I(t) \cdot J(x,y)(t), \tag{2.4}$$

where — notice that $1 - \alpha - \frac{1}{p} = \frac{1}{q} - \alpha$ and review (1.3) —

$$I(t) = \frac{1}{\Gamma(\alpha)} \left(\int_0^t (t-s)^{p(\alpha-1)} e^{ps} ds \right)^{\frac{1}{p}} = \frac{1}{\Gamma(\alpha)} \left(\int_0^t s^{p(\alpha-1)} e^{p(t-s)} ds \right)^{\frac{1}{p}}$$

$$= \frac{e^t}{\Gamma(\alpha)} \left[\int_0^{pt} \left(\frac{\tau}{p} \right)^{p(\alpha-1)} e^{-\tau} \cdot \frac{d\tau}{p} \right]^{\frac{1}{p}} \leq \frac{e^t}{\Gamma(\alpha)} \cdot p^{\frac{1}{q} - \alpha} \Gamma(1 - p(1-\alpha))$$

$$= c(\alpha, p) \cdot e^t$$

and — recall that $h(t) \geq 1$ —

$$J(x,y)(t) = \left\{ \int_0^t [F(s)]^q \cdot \frac{|x(s) - y(s)|^q}{e^{sq}} ds \right\}^{\frac{1}{q}}$$

$$= \left\{ \int_0^t \frac{d}{ds} \left\{ \frac{\exp \left(\lambda \int_0^s [h(\tau) F(\tau)]^q d\tau \right)}{\lambda} \right\} \left[\frac{|x(s) - y(s)|}{H_\lambda(s)} \right]^q ds \right\}^{\frac{1}{q}}$$

$$\leq \left\{ \int_0^t \frac{d}{ds} \left\{ \frac{\exp \left(\lambda \int_0^s [h(\tau) F(\tau)]^q d\tau \right)}{\lambda} \right\} ds \right\}^{\frac{1}{q}} \cdot d_\lambda(x,y)$$

$$\leq \frac{\exp \left(\frac{\lambda}{q} \int_0^t [h(\tau) F(\tau)]^q d\tau \right)}{\lambda^{\frac{1}{q}}} \cdot d_\lambda(x,y)$$

$$= \lambda^{-\frac{1}{q}} \cdot \frac{H_\lambda(t)}{e^t} \cdot d_\lambda(x,y).$$

By combining these estimates, we infer that

$$d_\lambda(Tx, Ty) \leq c(a,p)\lambda^{-\frac{1}{q}} \cdot d_\lambda(x,y) \quad \text{for all } x, y \in X.$$

The formula is valid only if we establish that $Tx \in X$ *whenever* $x \in X$. This follows from the next estimates

$$|(Tx)(t)| \leq |(Tx)(t) - (Tx_0)(t)| + |(Tx_0)(t)|$$

$$\leq c(a,p)\lambda^{-\frac{1}{q}} H_\lambda(t) \cdot d_\lambda(x, x_0) + h(t)$$

$$\leq H_\lambda(t) \cdot \left[c(a,p)\lambda^{-\frac{1}{q}} d_\lambda(x, x_0) + 1 \right]$$

$$= O(H_\lambda(t)) \quad \text{as } t \to +\infty. \tag{2.5}$$

In conclusion, the operator $T : X \to X$ is a contraction for every $\lambda > [c(a,p)]^q$. Its unique fixed point is the solution of (2.2) with (global) existence in the future. $\square$

The solution x of the integral counterpart of problem (2.1) which has been presented in Theorem 2.1 is usually called a *mild solution* of the problem. A natural question might be raised: given the impetus on the spaces $\mathcal{RL}^\beta$ in the preceding chapter, can one hope to find at least a solution $x \in C([0,+\infty),\mathbb{R}) \cap C^1((0,+\infty),\mathbb{R})$ with $t^\beta x' \in L^\infty_{\mathrm{loc}}([0,+\infty),\mathbb{R})$ for some $\beta \in [0,1)$ that verifies the fractional differential equation in (2.1) everywhere? Such functions are named frequently, for obvious reasons, *strong solutions* of (2.1).

To reply to this inquiry, allow us to introduce an auxiliary integral equation, namely

$$z(t) = z_0(t) + \int_0^t D(t,\tau)z(\tau)d\tau, \quad t \geq 0, \tag{2.6}$$

where $z_0 \in C(\mathbb{R}_+,\mathbb{R})$, $D \in C((0,+\infty)^2,\mathbb{R})$ and $D(t,\cdot) \in L^1((0,t),\mathbb{R})$ everywhere in $(0,+\infty)$.

First, noticing that the proof in Theorem 2.1 has been based on the definition of function Γ, we redesign that demonstration to allow more freedom for the functional coefficients.

Proposition 2.1. *Suppose $D(t,s) = D(t-s)$ and there exists $1 < p < +\infty$ such that the application $t \mapsto e^{-t} \cdot D(t)$ belongs to $L^p((0,+\infty),\mathbb{R})$. Then, the equation (2.6) has a unique solution in $C(\mathbb{R}_+,\mathbb{R})$.*

Proof. We shall start with the same "prospective" view of $z(t)$ as in Theorem 2.1. In fact, for small t's, the presumptive solution of (2.6) must stay "close" to $z_0(t)$ and so it makes sense to introduce the functions

$$h(t) = 1 + |z_0(t)| + \int_0^t |D(t-\tau) \cdot z_0(\tau)|d\tau$$

and

$$H_\lambda(t) = h(t) \cdot \exp\left(t + \frac{\lambda}{q}\int_0^t [h(s)]^q ds\right),$$

where $t > 0$, for a fixed $\lambda > 0$. Here, q is taken such that $\frac{1}{p} + \frac{1}{q} = 1$.

Recall the convolution technique from (1.12). By taking $F(x) = D(x) \cdot \chi_{(0,\frac{t}{2})}(x)$ and the same G — with $g = z_0$ —, we deduce that *the integral in the formula of h is an L^1_{loc}-function*. Via the change of variables $\tau = t - s$, we deduce that

$$\int_0^t |D(t-\tau) \cdot z_0(\tau)|d\tau = \int_0^t |D(s) \cdot z_0(t-s)|ds$$

$$\leq \int_0^t |D(s)|ds \cdot \|z_0\|_{L^\infty([0,t+1],\mathbb{R})}, \qquad (2.7)$$

see [Rudin (1987), Chap. 7, Th. 7.26]. This means that *the function h is also a member of* $L^\infty_{\text{loc}}(\mathbb{R}_+, \mathbb{R})$. The absolute continuity of the Lebesgue integral [Kolmogorov and Fomine (1974), p. 294, Th. 5] yields $\lim_{t\searrow 0} h(t) = 1 + |z_0(0)|$.

Further, take $t_2 > t_1 > 0$. We have

$$\left| \int_0^{t_2} D(s)z_0(t_2 - s)ds - \int_0^{t_1} D(s)z_0(t_1 - s)ds \right|$$

$$\leq \int_{t_1}^{t_2} |D(s) \cdot z_0(t_2 - s)|ds + \int_0^{t_1} |D(s) \cdot [z_0(t_2 - s) - z_0(t_1 - s)]|\, ds$$

$$\leq \int_{[t_1,t_2]} |D(s)|ds \cdot \|z_0\|_{L^\infty([0,1+t_2],\mathbb{R})}$$

$$+ \|z_0(t_2 - \cdot) - z_0(t_1 - \cdot)\|_{L^\infty([0,1+t_2],\mathbb{R})} \cdot \|D\|_{L^1((0,1+t_2),\mathbb{R})}$$

and, taking into account both the absolute continuity of the Lebesgue integral (of D) and the local uniform continuity of the continuous function z_0, conclude that the mapping $t \mapsto \int_0^t D(s)z_0(t - s)ds$ is actually in $C((0, +\infty), \mathbb{R})$.

So, $h \in C((0, +\infty), \mathbb{R})$ and, since $\lim_{t\searrow 0} h(t) \in (0, +\infty)$, we deduce that *the function h has an extension* $\tilde{h} \in C(\mathbb{R}_+, \mathbb{R})$. For simplicity, we shall write h instead of $\tilde{h}$ from now on.

Observe that, given $\xi > 0$, one has $e^{-p\xi} \cdot \int_0^\xi |D(s)|^p ds \leq \int_0^\xi \left[\frac{|D(s)|}{e^s} \right]^p ds < +\infty$. Consequently, $D \in L^p_{\text{loc}}([0, +\infty), \mathbb{R})$.

The *continuity of* H_λ *in* $\mathbb{R}_+$ is an immediate consequence of the previous inferences.

Further, introduce the complete metric space $\mathcal{Z} = (Z, d_\lambda)$, where Z is the set of all functions $z \in C(\mathbb{R}_+, \mathbb{R})$ with $z(t) = O(H_\lambda(t))$ when $t \to +\infty$ while the distance d_λ is given by (2.3). The operator $T : \mathcal{Z} \to C(\mathbb{R}_+, \mathbb{R})$ is defined by means of the right-hand part of (2.6).

The homologous of (2.4) reads as

$$|(Tz_1)(t) - (Tz_2)(t)| \leq \left[\int_0^t |D(t - \tau)|^p d\tau \right]^{\frac{1}{p}} \cdot \left[\int_0^t |z_1(\tau) - z_2(\tau)|^q \right]^{\frac{1}{q}}$$

$$\leq \left[\int_0^t |D(t - \tau) \cdot e^\tau|^p \, d\tau \right]^{\frac{1}{p}}$$

$$\times \left\{ \int_0^t \frac{d}{d\tau} \left\{ \frac{\exp\left(\lambda \int_0^\tau [h(s)]^q ds \right)}{\lambda} \right\} \right\}^{\frac{1}{q}} \cdot d_\lambda(z_1, z_2)$$

$$\leq e^t \left[\int_0^t |D(\tau) \cdot e^{-\tau}|^p \, d\tau \right]^{\frac{1}{p}} \cdot \frac{H_\lambda(t)}{\lambda^{\frac{1}{q}} e^t} \, d_\lambda(z_1, z_2)$$

and it suffices to take $\lambda > \left[\int_0^{+\infty} |D(\tau) e^{-\tau}|^p \, d\tau \right]^{\frac{1}{p-1}}$ to obtain a contraction in $\mathcal{Z}$. The essential estimate (2.5) is obtained by taking $c(\alpha, p) = c(p) = \|e^{-t} D\|_{L^p((0,+\infty),\mathbb{R})}$ there. $\square$

Second, notice that for $D(t, \tau) = \frac{\chi_{(0,+\infty)}(t-\tau)}{B(\alpha, 1-\alpha)} \cdot \tau^{-\alpha}(t-\tau)^{1-\alpha}$, where $\alpha \in (0,1)$ and $z_0 = 0$, the integral equation (2.6) has an infinity of solutions in $C^1(\mathbb{R}_+, \mathbb{R})$, namely *all the constant functions*. This means various restrictions must be imposed on the kernel $D(t, \tau)$ to ensure *uniqueness* of solution to (2.1) even in the case of mild solutions. The particular circumstances when $D(t, \tau) = D_1(\tau) \cdot (t-\tau)^{-\beta} \chi_{(0,+\infty)}(t-\tau)$ and respectively $D(t, \tau) = D(t-\tau)\chi_{(0,+\infty)}(t-\tau)$ have been given some sufficient conditions for uniqueness in the preceding results.

Third, let us see what it means having a smooth solution of (2.2). Take, not to complicate matters, $f(t, x) = a(t)x$, where $a \in C(\mathbb{R}_+, \mathbb{R})$. Now, for $y(t) = x'(t)$, we get

$$\int_0^t y(s) ds = \frac{1}{\Gamma(\alpha)} \int_0^t \frac{a(s)}{(t-s)^{1-\alpha}} \cdot \left[x_0 + \int_0^s y(\tau) d\tau \right] ds$$

$$= \frac{x_0}{\Gamma(\alpha)} \int_0^t \frac{a(s)}{(t-s)^{1-\alpha}} ds + \frac{1}{\Gamma(\alpha)} \int_{[0,t]^2} \frac{a(s)y(\tau)}{(t-s)^{1-\alpha}}$$
$$\times \chi_{(0,+\infty)}(s-\tau) d\tau ds$$

$$= \frac{x_0}{\Gamma(\alpha)} \int_0^t \frac{a(s)}{(t-s)^{1-\alpha}} ds + \frac{1}{\Gamma(\alpha)} \int_0^t \int_{\{s > \tau\}} \frac{a(s)y(\tau)}{(t-s)^{1-\alpha}} ds d\tau$$

$$= \frac{x_0}{\Gamma(\alpha)} \int_0^t \frac{a(s)}{(t-s)^{1-\alpha}} ds + \frac{1}{\Gamma(\alpha)} \int_0^t \int_\tau^t \frac{a(s)y(\tau)}{(t-s)^{1-\alpha}} ds d\tau$$

$$= x_0 \cdot b(t,0) + \int_0^t y(\tau) \cdot b(t, \tau) d\tau,$$

where $b(t, \tau) = \frac{1}{\Gamma(\alpha)} \int_\tau^t \frac{a(s)}{(t-s)^{1-\alpha}} ds \cdot \chi_{(0,+\infty)}(t-\tau)$.

Suppose, along this line of thinking, that we can differentiate everywhere the previous identity, thus getting to

$$y(t) = x_0 \cdot c(t,0) + \int_0^t c(t, \tau) y(\tau) d\tau, \quad t > 0,$$

for $c(t, \tau) = \frac{\partial b}{\partial t}(t, \tau)$. Moreover, let's consider that $y \in \mathcal{RL}^\alpha$, and so,

$$z(t) = x_0 c(t,0) t^\alpha + t^\alpha \int_0^t \frac{c(t, \tau)}{\tau^\alpha} \cdot z(\tau) d\tau, \quad z(t) = t^\alpha \cdot y(t), \quad (2.8)$$

everywhere in $\mathbb{R}_+$. Admitting that $\lim_{t\searrow 0}[t^\alpha c(t,0)] \in \mathbb{R}$, the latter integral equation is exactly (2.6). Here, $D(t,\tau) = c(t,\tau) \cdot \left(\frac{t}{\tau}\right)^\alpha \chi_{(0,+\infty)}(\tau)$.

Now, to give an answer to the inquiry from page 30, we say: *yes*, at least *locally*, we can find strong solutions to the fractional differential equations with a Caputo operator of order $\alpha \in (0,1)$. And we base our statement on the next result.

Theorem 2.2. *The initial value problem (2.1), where $\alpha \in \left(\frac{1}{2},1\right)$ and $f(t,x) = A \cdot x$ for some $A \in \mathbb{R}$, has a unique continuously differentiable solution x, with $x' \in \mathcal{RL}^\alpha$, in small interval to the right of 0.*

Proof. Set $a(t) = A$ in (2.8). We have — recall (1.3) and take $t > \tau$ —

$$b(t,\tau) = \frac{A}{\Gamma(1+\alpha)} \cdot (t-\tau)^\alpha, \quad c(t,\tau) = \frac{A}{\Gamma(\alpha)} \cdot \frac{1}{(t-\tau)^{1-\alpha}},$$

and

$$D(t,\tau) = \frac{A}{\Gamma(\alpha)} \cdot \frac{t^\alpha}{\tau^\alpha(t-\tau)^{1-\alpha}}. \tag{2.9}$$

Introduce the function $f \in C((0,+\infty),\mathbb{R}) \cap L^1_{\text{loc}}(\mathbb{R}_+,\mathbb{R})$ with the formula $f(t) = \frac{A}{\Gamma(\alpha)} \cdot t^{-\alpha}\chi_{(0,+\infty)}(t)$. Observe as well that $f \in \mathcal{RL}^\alpha$. Also, it can be established straightforwardly that $\int_0^t D(t,\tau)d\tau = t^\alpha \cdot (Q_{1-\alpha,1}f)(t) = A\Gamma(1-\alpha) \cdot t^\alpha$, see (1.30), when $t > 0$.

Recall (1.35). By means of the change of variables $\tau = 1 - \lambda$, we notice that, if $\beta = 1 - \alpha$, $\gamma = \alpha$ and $a = 0$,

$$\int_{t_1}^{t_2} \frac{ds}{(t_2-s)^\beta(s-a)^\gamma} = \int_{\frac{t_1-a}{t_2-a}}^1 \frac{d\lambda}{(1-\lambda)^\beta\lambda^\gamma}$$

$$= \int_0^{\frac{t_2-t_1}{t_2-a}} \frac{d\tau}{\tau^\beta(1-\tau)^\gamma}, \quad t_2 > t_1 > a.$$

Taking into account the absolute continuity of the Lebesgue integral , we conclude that, as a function of t_1, t_2, the latter integral is continuous in $(0,+\infty)^2$. Similarly, via (1.40), (1.43), (1.46), we establish that the mapping $\mathfrak{D}$ given by $t \mapsto \int_0^t D(t,\tau)z(\tau)d\tau$ is continuous in $(0,+\infty)$, where D has been displayed in (2.9) and $z \in C(\mathbb{R}_+,\mathbb{R})$.

Further, consider the Banach space $\mathcal{Z} = (C([0,T],\mathbb{R}), \|\cdot\|_\infty)$, where $T > 0$ is taken such that $k = T^\alpha \cdot A\Gamma(1-\alpha) < 1$. Then, we claim that the operator $\mathcal{T} : \mathcal{Z} \to \mathcal{Z}$ with the formula

$$\mathcal{T}(z)(t) = z_0(t) + \int_0^t D(t,\tau)z(\tau)d\tau, \quad t \in [0,T], z \in \mathcal{Z},$$

where $z_0(t) = x_0 c(t,0)t^\alpha = X_0 \cdot t^{2\alpha-1}$, $X_0 = \frac{Ax_0}{\Gamma(\alpha)}$, *is well-defined,* meaning $T(z)$ is continuous everywhere in $[0,T]$. To validate the assertion, remark firstly that the restriction on α has been imposed to ensure that $z_0 \in \mathcal{Z}$. Second, the estimate (2.7) is obviously correct and thus $\lim_{t\searrow 0} \left| \int_0^t D(t,\tau)z(\tau)d\tau \right| = \lim_{t\searrow 0}[t^\alpha \cdot A\Gamma(1-\alpha)\|z\|_{L^\infty([0,T],\mathbb{R})}] = 0$ for all $z \in \mathcal{Z}$. Since the mapping $\mathfrak{D}$ can now be continuated in a continuous manner downward to $t = 0$, we conclude that the claim is truthful.

The estimates

$$|T(z_1)(t) - T(z_2)(t)| \le \int_0^t |D(t,\tau)| \cdot |z_1(\tau) - z_2(\tau)|d\tau$$

$$\le \int_0^t |D(t,\tau)|d\tau \cdot \|z_1 - z_2\|_{L^\infty([0,T],\mathbb{R})}$$

$$\le k \cdot \|z_1 - z_2\|_{L^\infty([0,T],\mathbb{R})}, \quad t \in [0,T],$$

yield the existence of a unique solution z_* of (2.6) in $\mathcal{Z}$, which is the fixed point of the contraction T.

The function $x \in C([0,T],\mathbb{R}) \cap C^1((0,T],\mathbb{R})$ given by

$$x(t) = x_0 + \int_0^t \frac{z_*(s)}{s^\alpha}ds, \quad 0 \le t \le T,$$

is the strong solution of the problem (2.1) we have been looking for. $\square$

2.1   A Lovelady-Martin Uniqueness Result for the Equation (2.2)

Consider the ordinary differential equation

$$x'(t) = f(x(t)), \quad t \ge 0, \tag{2.10}$$

for a continuous function $f : \mathbb{R} \to \mathbb{R}$.

Typically, when the nonlinearity f of the equation is not of Lipschitz type [Agarwal and Lakshmikantham (1993)], there are only a few techniques to help us decide whether an initial value problem attached to it has more than one solution. As an example, the equation (2.10) for $f(x) = \sqrt{x} \cdot \chi_{(0,+\infty)}(x)$ has an infinity of solutions $(x_T)_{T\ge 0}$, with $x_T(t) = \frac{(t-T)^2}{4} \cdot \chi_{(T,+\infty)}(t)$, defined on the nonnegative half-line which start from $x(0) = 0$.

On the other hand, one may wonder naturally whether the solutions x_T's are *the only ones that are not identically null*. We recall that an initial value problem with two non-trivial solutions must have a *continuum* [Whyburn (1964), p. 12] of solutions by means of H. Kneser's theorem [Hartman (1964), p. 15]. In such a simple case, however, the answer is affirmative: there are no other non-trivial solutions. This follows from the fact that $f(x)$ is locally Lipschitz throughout $(0, +\infty)$ since, if there are two solutions $x(t)$, $y(t)$ starting from zero and meeting each other for the second time at $t_0 > 0$, where $x(t_0) = y(t_0) = x_0 > 0$, then in a small interval of positive numbers centered at t_0 they should coincide. Such a conclusion, obviously, will contradict the minimality of t_0.

Another demonstration of the previous claim follows from the Lovelady-Martin uniqueness theorem [Lovelady and Martin (1972); Lovelady (1973)]. In fact, notice that the nonlinearity $f(x)$ verifies the inequality

$$
\begin{aligned}
\left| x - y - c\left(\sqrt{x} - \sqrt{y}\right) \right| &= \left[c - \left(\sqrt{x} + \sqrt{y}\right) \right] \cdot \left| \sqrt{x} - \sqrt{y} \right| \\
&= \frac{c + o(1)}{\sqrt{x} + \sqrt{y}} \cdot |x - y| \\
&> |x - y|
\end{aligned}
\tag{2.11}
$$

in a small interval to the right of 0 for a fixed $c > 0$. The estimate (2.11) is of *Lovelady-Martin type* and ensures that the initial value problem will have at most one solution which reaches x_0 in finite time for every $x_0 > 0$.

Our intention in this section is to establish a variant of the Lovelady-Martin criterion of uniqueness in the case of fractional differential equations with a Caputo derivative. Before giving its statement, however, allow us to demonstrate several auxiliary results, of a rather classical nature [McShane (1950)], from which we shall benefit in the course of proving the uniqueness criterion.

Given the Banach space $(B, \| \cdot \|)$, set $u, v \in B$ and introduce the functions $\xi, \zeta : \mathbb{R}\backslash\{0\} \to \mathbb{R}$ with the formulas

$$
\xi(\varepsilon) = \frac{\|u + \varepsilon \cdot v\| - \|u\|}{\varepsilon} = \frac{\|u\| - \|u - \eta \cdot v\|}{\eta} = \zeta(\eta), \quad \varepsilon = -\eta \neq 0.
$$

Lemma 2.1. *ξ is monotone nondecreasing, ζ is monotone nonincreasing, both of them are bounded and*

$$
|\xi(\varepsilon)| = |\zeta(-\varepsilon)| \leq \|v\|, \quad \varepsilon \neq 0.
$$

Proof. Take $\varepsilon > \delta > 0$ and deduce via the triangle inequality that

$$
\begin{aligned}
\xi(\delta) - \xi(\varepsilon) &\leq (\varepsilon\delta)^{-1} \left[\|\varepsilon\|u + \delta v\| - \delta\|u + \varepsilon v\|\| - (\varepsilon - \delta)\|u\| \right] \\
&= (\varepsilon\delta)^{-1} \left[\|\|\varepsilon u + \varepsilon\delta v\| - \| - \delta u - \delta\varepsilon v\|\| - (\varepsilon - \delta)\|u\| \right]
\end{aligned}
$$

$$\leq \frac{\|(\varepsilon - \delta)u\| - (\varepsilon - \delta)\|u\|}{\varepsilon\delta} = 0.$$

Thus, the function ξ is monotone nondecreasing on the positive half-line. The monotonicity on the negative half-line is proved verbatim. The conclusion regarding ζ follows from the monotonicity of ξ by taking into account the identity $\zeta(\varepsilon) = \xi(-\varepsilon)$.

Now, set ε, η in $(0, +\infty)$. Thus, we have, appealing again to the triangle inequality,

$$\frac{\|u\| - \|u - \eta \cdot v\|}{\eta} - \frac{\|u + \varepsilon \cdot v\| - \|u\|}{\varepsilon}$$

$$= \frac{(\varepsilon + \eta)\|u\| - (\varepsilon\|u - \eta v\| + \eta\|u + \varepsilon v\|)}{\varepsilon\eta}$$

$$\leq \frac{(\varepsilon + \eta)\|u\| - \|(\varepsilon + \eta)u\|}{\varepsilon\eta} = 0,$$

which means that $\xi(\tau_1) \leq \xi(\tau_2)$ for all $\tau_1 < 0 < \tau_2$. $\square$

Given the boundedness of ξ, we deduce also that

$$\lim_{\varepsilon \searrow 0} \xi(\varepsilon) = \inf_{\varepsilon \searrow 0} \frac{\|u + \varepsilon v\| - \|u\|}{\varepsilon}, \quad \lim_{\varepsilon \nearrow 0} \xi(\varepsilon) = \inf_{\eta \searrow 0} \frac{\|u\| - \|u - \eta v\|}{\eta}.$$

Further, assume that the function $x : \mathbb{R} \to B$ is differentiable at some point $t \in \mathbb{R}$ and take $k(s) = \|x(s)\|$, where $s \in \mathbb{R}$.

Lemma 2.2. *The following identities hold true:*

$$k_r'(t) = \lim_{h \searrow 0} \frac{\|x(t + h)\| - \|x(t)\|}{h} = \lim_{h \searrow 0} \frac{\|x(t) + h \cdot x'(t)\| - \|x(t)\|}{h}, \quad (2.12)$$

and respectively

$$k_\ell'(t) = \lim_{h \searrow 0} \frac{\|x(t)\| - \|x(t - h)\|}{h} = \lim_{h \searrow 0} \frac{\|x(t)\| - \|x(t) - h \cdot x'(t)\|}{h}. \quad (2.13)$$

Also, $-\|x'(t)\| \leq k_\ell'(t) \leq k_r'(t) \leq \|x'(t)\|$.

Proof. The limits on the right-hand side of the identities exist thanks to the monotonicity properties of ξ for $u = x(t)$ and $v = x'(t)$.

Now, we have

$$\frac{\|x(t + h)\| - \|x(t)\|}{h} = \frac{\|x(t) + hx'(t) + g(h)\| - \|x(t)\|}{h}$$

$$\leq \frac{\|x(t) + hx'(t)\| + \|g(h)\| - \|x(t)\|}{h}$$

$$= \frac{\|x(t) + hx'(t)\| - \|x(t)\|}{h} + o(1), \quad h \searrow 0,$$

where $g(h) = o(|h|)$ for $h \to 0$, which leads to

$$\limsup_{h \searrow 0} \frac{\|x(t+h)\| - \|x(t)\|}{h} \leq \lim_{h \searrow 0} \frac{\|x(t) + h \cdot x'(t)\| - \|x(t)\|}{h}. \quad (2.14)$$

Analogously,

$$\frac{\|x(t+h)\| - \|x(t)\|}{h} \geq \frac{\|x(t) + hx'(t)\| - \|g(h)\| - \|x(t)\|}{h}$$

$$= \frac{\|x(t) + hx'(t)\| - \|x(t)\|}{h} + o(1), \quad h \searrow 0,$$

and so

$$\liminf_{h \searrow 0} \frac{\|x(t+h)\| - \|x(t)\|}{h} \geq \lim_{h \searrow 0} \frac{\|x(t) + h \cdot x'(t)\| - \|x(t)\|}{h}. \quad (2.15)$$

The conclusion (2.12) follows from (2.14), (2.15). The second part (2.13) is established verbatim. The last (triple) inequality in the statement is a consequence of the monotonicity properties of ξ. $\square$

We recall at this point that, in many not so uncommon situations, the integral equation (2.2) possesses C^1–solutions, see Theorem 2.2. To emphasize this with help from a simple example, consider the initial value problem

$$\begin{cases} {}_0C_t^\alpha(x)(t) = c \cdot [x(t)]^\beta, \, t > 0, \\ x(0) = 0, \end{cases} \quad (2.16)$$

where $\beta \in \left(0, \frac{1}{3}\right)$ is fixed, $\alpha + \beta > 1$ and[1]

$$c = \frac{\Gamma(\alpha)}{B(1 + (1+\varepsilon)\beta, \alpha)} = \frac{\Gamma(1 + \alpha + (1+\varepsilon)\beta)}{\Gamma(1 + (1+\varepsilon)\beta)} = \frac{\Gamma(2+\varepsilon)}{\Gamma(2+\varepsilon-\alpha)},$$

where $\varepsilon = \frac{\alpha}{1-\beta} - 1 > 0$. A nontrivial C^1–solution of (2.16) is provided by $x(t) = t^{1+\varepsilon}$, where $t \geq 0$.

Theorem 2.3. ([Baleanu et al. (2013), Theorem 1]) *Let $X \subseteq C^1([0,T], \mathbb{R})$ be a non-void family of functions, with $T > 0$ arbitrarily fixed. Assume that there exist two numbers $c > 0$ and $C \in (0,1)$ such that the inequality*

$$\left| x(t) - y(t) - c \cdot {}_0D_t^{1-\alpha}[f(t,x) - f(t,y)](t) \right| \geq C \cdot |x(t) - y(t)|$$

holds almost everywhere with respect to the Lebesgue measure m on $[0,T]$ for every pair $(x,y) \in X^2$. Here, $f(t,x)$ designates the mapping $t \mapsto f(t,x(t))$, where $x \in X$.

Then, the integral equation (2.2) has at most one solution in X.

---

[1] Via (1.3).

Proof. Let x, $y \in X$ be solutions of (2.2). As the absolute value of the real numbers is a norm in $\mathbb{R}$, the function k given by $k(t) = |x(t) - y(t)|$ for any $t \in [0, T]$ is differentiable to the left everywhere in $(0, T)$. So, we may employ (2.13).

Now, there exists a Lebesgue measurable set $E \subset [0, T]$, with $m(E) = 0$, such that

$$
\begin{aligned}
k'_\ell(t) &= \lim_{h \searrow 0} \frac{1}{h} \left\{ |x(t) - y(t)| - |x(t) - y(t) - h[x'(t) - y'(t)]| \right\} \\
&= \lim_{h \searrow 0} \frac{1}{h} \left\{ |x(t) - y(t)| - \left| x(t) - y(t) - h \cdot {}_0D_t^{1-\alpha}[f(t, x) - f(t, y)](t) \right| \right\} \\
&\leq \frac{1}{c} \left\{ |x(t) - y(t)| - \left| x(t) - y(t) - c \cdot {}_0D_t^{1-\alpha}[f(t, x) - f(t, y)](t) \right| \right\} \\
&\leq \eta \cdot |x(t) - y(t)| \\
&= \eta k(t), \quad t \in (0, T] \backslash E, \quad \text{where } \eta = \frac{1 - C}{c}.
\end{aligned}
\tag{2.17}
$$

Further, set $\alpha > 0$ and $q(t) = \alpha \cdot e^{\eta t} + \frac{\alpha}{\eta}(e^{\eta t} - 1)$ for $t \in [0, T]$. Obviously, $q'(t) = \eta \cdot q(t) + \alpha > \eta q(t)$ everywhere. We have $k(0) = 0 < \alpha = q(0)$ and so, on a small interval to the right of 0, we get $0 \leq k(t) < q(t)$.

We claim that *the double inequality* $0 \leq k(t) \leq q(t)$ *holds everywhere in* $[0, T]$. To prove the assertion, suppose for the sake of contradiction that there exist some numbers $0 < t_0 < t_1 < T$ such that

$$
\begin{cases}
k(t) \leq q(t), \, t \in [0, t_0), \\
k(t) = q(t), \, t = t_0, \\
k(t) > q(t), \, t \in [t_0, t_1].
\end{cases}
$$

The main difficulty here is to decide what to do when $t_0 \in E$. First, take a point $t_E \in E$. According to the corollary of [Rudin (1987), Chapter 2, Theorem 2.22], every Lebesgue-measurable set with a positive measure possesses a subset which is not Lebesgue measurable. In particular, the subset is not void as the void set is of measure zero. So, by taking the sequence of open intervals $I_n = \left(t_E - \frac{1}{n}, t_E + \frac{1}{n} \right)$ for all the n's large enough to have $I_n \subset [0, T]$, we deduce that there exists a sequence of real numbers $(t_n)_n$ such that $t_n \in I_n \backslash E$. Obviously, $\lim_{n \to +\infty} t_n = t_E$ and $k'_\ell(t_n) \leq (\eta \cdot k)(t_n)$. If we can establish somehow that k'_ℓ is *continuous at* $t_E = t_0$ then, by letting $n \to +\infty$ in the latter inequality, we deduce that the key estimate (2.17) is valid in t_0.

To keep things as general as possible, we assume that $k'_\ell(t_0) < 0$ without carrying about this continuity. Using the definition of k'_ℓ, we infer that there exists a small interval to the left of t_0 where the values of k are greater that

$k(t_0) = q(t_0)$. However, since q increases everywhere, in the same interval the values of q are all less that $q(t_0)$. This means that q lies below k, which contradicts the definition of t_0.

Further, suppose that $k'_\ell(t_0) = 0$ and set $\varepsilon > 0$ of small size. There exists $h_\varepsilon > 0$, with $[t_0 - h_\varepsilon, t_0] \subset [0, T]$, such that $k(t_0) - \varepsilon \cdot h \le k(t_0 - h)$ when $h \in [0, h_\varepsilon]$. For the same h, we have $q(t_0 - h) = q(t_0) + q'(t_0) \cdot (-h) + o(h) = k(t_0) - [\eta q(t_0) + \alpha + o(1)]h$. By asking that $\varepsilon < \frac{\eta\alpha}{2} \le \frac{\eta q(t_0)}{2}$, we arrive again at a contradiction: $q(t_0 - h) < k(t_0 - h)$.

It remains to settle the essential case $k'_\ell(t_0) > 0$. Recalling the definition of k as well as the last statement of Lemma 2.2, we have $|(x - y)'(t_0)| \ge k'_\ell(t_0) > 0$ and also $|(x - y)(t_0)| = q(t_0) > 0$. This means that, in a small interval centered at t_0, *the non-null sign of the function* $x - y$ *does not change*. So, the function k is continuously differentiable around t_0, and, in particular, k'_ℓ is continuous at t_0. The inequality (2.17) now holds true at t_0. But since the (semi-) slope of q there is greater than the (semi-) slope of k, we get to a contradiction (as before, there would be a small interval to the left of t_0 where q lies below k).

We have obtained that $0 \le k(t) \le q(t)$ for all $t \in [0, T]$. By making $\alpha \searrow 0$, we conclude that $k(t) = 0$ throughout $[0, T]$, which means that x and y coincide to the right of 0. The proof is complete. $\square$

2.2   A Nagumo-like Uniqueness Criterion for the Fractional Differential Equations with a Riemann-Liouville Derivative

As before, the starting point for our investigation is the rich field of research regarding the *uniqueness of solution* for various families of ordinary differential equations which resemble (2.10). In fact, let us consider the ordinary differential equation

$$x'(t) + f(t, x(t)) = 0, \quad t > 0, \tag{2.18}$$

where the function $f : (0, +\infty) \times \mathbb{R} \to \mathbb{R}$ has been presumed continuous.

It has been established by Nagumo [Nagumo (1926)] that, if $\lim_{t \searrow 0} f(t, x) = 0$ uniformly with respect to $x \in [-1, 1]$ and also

$$|f(t, x)| \le \frac{|x|}{t}, \quad x \in \mathbb{R}, \ t > 0, \tag{2.19}$$

the only solution of the equation (2.18) starting from $x(0) = 0$ is the identically null solution.

To convince yourself quickly of this fact, just take $f(x) = x$ in (2.10) and observe that the inequality (2.19) is valid for all the pairs $(t, x) \in (0, \frac{1}{2}] \times \mathbb{R}$. Obviously, the only solution departing from zero of a homogenous first-order linear differential equation remains null for as long as it exists.

For a thorough review of the literature concerning the Nagumo theorem and several important generalizations, we recommend the paper by Athanassov [Athanassov (1990)]. Recently, see [Constantin (2010)], the inequality regarding f has been replaced by

$$|f(t, x)| \leq \frac{\omega(|x|)}{t}, \quad \text{where} \quad \int_0^r \frac{\omega(s)}{s}\, ds \leq r, \, r \geq 0,$$

and $\omega : [0, +\infty) \to [0, +\infty)$ is a continuous, increasing function, with $\omega(0) = 0$. The latter result of Constantin has been extended to the case of the second order ordinary differential equations in [Mustafa (2012)]. For Nagumo-like results devised for higher order equations, see [Nagumo (1927); Wintner (1956); Constantin (1996)].

To produce a fractional calculus variant for Constantin's extension of the classical Nagumo uniqueness criterion, one has to understand first the mechanics behind Nagumo's idea.

Given x a solution to the equation (2.18) such that $x(0) = 0$ and assuming it is not trivial, the core of Nagumo's approach to uniqueness is to reach a contradiction by *prescribing* the asymptotic behavior of the solution. To be precise, *set $\alpha \in (0, 1]$ and assume that $x(t) = o(t^{\alpha})$ as $t \searrow 0$*. Originally, $\alpha = 1$, see the comments from [Athanassov (1990); Wintner (1956); Constantin (2010); Mustafa and O'Regan (2011)].

Via the L'Hôpital rule, to get this type of asymptotic development for any solution of (2.18) it suffices having $\lim_{t \searrow 0} \frac{x'(t)}{\alpha t^{\alpha-1}} = 0$, which gives us the first hypothesis on f, namely *we assume that $\lim_{t \searrow 0} t^{1-\alpha} f(t, x) = 0$ uniformly with respect to $x \in [-1, 1]$*. A second hypothesis is needed, this being an inequality for f:

$$|f(t, x)| \leq \alpha \cdot \frac{|x|}{t}, \quad x \in \mathbb{R}, \, t > 0.$$

Now, introduce the continuous function $y_{\alpha} : [0, T] \to [0, +\infty)$ by means of the formula

$$y_{\alpha}(t) = \begin{cases} \sup_{s \in (0, t]} \frac{|x(s)|}{s^{\alpha}}, & t \in (0, T], \\ 0, & \text{otherwise,} \end{cases} \tag{2.20}$$

for some $T > 0$ in the domain of existence of x. By integrating (2.18), we get

$$|x(t)| \leq \int_0^t \alpha \frac{|x(s)|}{s} ds = \int_0^t \frac{|x(s)|}{s^\alpha} \cdot \alpha \frac{ds}{s^{1-\alpha}}$$

$$< \varepsilon_T \cdot \int_0^t \alpha \frac{ds}{s^{1-\alpha}} = \varepsilon_T t^\alpha, \quad t \in (0, T]. \tag{2.21}$$

Here, $\varepsilon_T = y_\alpha(T) = \frac{|x(\xi_{T,\alpha})|}{\xi_{T,\alpha}^\alpha}$ for some $\xi_{T,\alpha} \in (0, T]$.

So,

$$\frac{|x(t)|}{t^\alpha} < \varepsilon_T, \quad t \in (0, T].$$

Taking $t = \xi_{T,\alpha}$ in the latter inequality, we arrive at a contradiction.

In the limiting case of $\alpha = 1$, it has been demonstrated using counterexamples that *the constant 1 in front of the ratio from the inequality (2.19) concerning f is optimal*, see [Athanassov (1990)]. This feature has also to be taken into consideration by any "fractional" variant of the Nagumo uniqueness theorem.

Fortunately enough, when computing the integral representation of a fractional differential equation with a Riemann-Lioville derivative (1.5), recall (1.55), one must *prescribe* the behavior of its solutions in the vicinity of the starting time 0. That is, for $\alpha \in (0, 1)$, given the initial value problem

$$\begin{cases} {}_0D_t^\alpha(x)(t) + f(t, x(t)) = 0, \, t > 0, \\ \lim_{t \searrow 0}[t^{1-\alpha} x(t)] = x_0 \in \mathbb{R}, \end{cases} \tag{2.22}$$

where the function $f : \mathbb{R}_+ \to \mathbb{R}$ is assumed continuous and $f(t, 0) = 0$ everywhere in $\mathbb{R}_+$, we have the integral equation homologous to (1.58)

$$x(t) = \frac{x_0}{t^{1-\alpha}} - \frac{1}{\Gamma(\alpha)} \int_0^t \frac{f(s, x(s))}{(t-s)^{1-\alpha}} ds, \quad t > 0. \tag{2.23}$$

Again, any solution $x \in C((0, +\infty), \mathbb{R})$ of the equation (2.23) is oft-called a *mild solution* of the problem (2.22). Our fractional variant of the Nagumo theorem is concerned only with these mild solutions.

Proposition 2.2. *Assume that the nonlinearity f satisfies the inequality*

$$|f(t, x)| \leq k \cdot \frac{|x|}{t^\delta}, \quad |x| \leq 1, t > 0, \tag{2.24}$$

for some $k > 0$ and $\delta \in (0, \alpha)$. Then, the only mild solution of (2.22) with $x_0 = 0$ is the identically null solution.

Proof. Suppose, for the sake of contradiction, that there exists the nontrivial solution x defined on $(0, X)$, where $X \leq +\infty$. Fix $T \in (0, X)$ small enough to have

$$\frac{k}{\Gamma(\alpha)} \cdot T^{\alpha-\delta} \cdot \int_0^1 \frac{d\lambda}{(1-\lambda)^{1-\alpha}\lambda^{1-\alpha+\delta}} < 1. \qquad (2.25)$$

Notice that $1 - \alpha + \delta < 1$, which means that the previous integral can be expressed in terms of the Beta function[2], see (1.3).

Introduce the continuous function $z_\alpha : [0, T] \to [0, +\infty)$ via the formula $z_\alpha = y_{\alpha-1}$, where y has been defined in (2.20).

The equation (2.23) leads to

$$|x(t)| \leq \frac{k}{\Gamma(\alpha)} \int_0^t \frac{s^{1-\alpha}|x(s)|}{(t-s)^{1-\alpha}} \cdot \frac{ds}{s^{1-\alpha+\delta}} < \varepsilon_T \cdot \frac{k}{\Gamma(\alpha)} \int_0^t \frac{ds}{(t-s)^{1-\alpha}s^{1-\alpha+\delta}}$$

$$= \frac{\varepsilon_T}{t^{1-\alpha}} \cdot \frac{k}{\Gamma(\alpha)} \cdot t^{\alpha-\delta} \cdot \int_0^1 \frac{d\lambda}{(1-\lambda)^{1-\alpha}\lambda^{1-\alpha+\delta}}$$

and respectively

$$t^{1-\alpha}|x(t)| \leq z_\alpha(t) \qquad (2.26)$$

$$\leq \frac{k}{\Gamma(\alpha)} \cdot T^{\alpha-\delta} \cdot \int_0^1 \frac{d\lambda}{(1-\lambda)^{1-\alpha}\lambda^{1-\alpha+\delta}} \cdot \varepsilon_T, \quad t \in (0, T],$$

where $\varepsilon_T = z_\alpha(T) = y_{\alpha-1}(T) = \xi_{T,\alpha-1}^{1-\alpha} \cdot |x(\xi_{T,\alpha-1})|$.

Taking $t = \xi_{T,\alpha-1}$ in the preceding inequality, we arrive at a contradiction.

So, on the very short subinterval $(0, T]$ of the domain of existence $(0, X)$ the solution is identically null. But *can it become nontrivial later on?* Suppose for the sake of contradiction that this is the case. Set the number $\Delta \in (0, X)$ such that

$$\frac{k}{\Gamma(1+\alpha)} \cdot \frac{\Delta^\alpha}{T^\delta} < 1,$$

where $T > 0$ has been introduced in (2.25).

Returning to the equation (2.23), we deduce that

$$|x(t)| \leq \frac{1}{\Gamma(\alpha)} \int_T^t \frac{|f(s, x(s))|}{(t-s)^{1-\alpha}} ds \leq \frac{k}{\Gamma(\alpha)} \int_T^t \frac{|x(s)|}{(t-s)^{1-\alpha}s^\delta} ds$$

$$\leq \frac{k}{\Gamma(\alpha)} \cdot \frac{1}{T^\delta} \cdot \frac{(t-T)^\alpha}{\alpha} \cdot \|x\|_{L^\infty([T,T+\Delta],\mathbb{R})}, \quad t \in [T, T+\Delta],$$

---

[2]It is, in fact, equal to $B(\alpha - \delta, \alpha)$.

and get to a contradiction again

$$\|x\|_{L^\infty([T,T+\Delta],\mathbb{R})} \leq \frac{k}{\Gamma(\alpha)} \cdot \frac{1}{T^\delta} \cdot \frac{\Delta^\alpha}{\alpha} \cdot \|x\|_{L^\infty([T,T+\Delta],\mathbb{R})}$$

$$< \|x\|_{L^\infty([T,T+\Delta],\mathbb{R})}. \tag{2.27}$$

So, *the solution is identically null* $[T, T + \Delta]$. Given that the interval $[T, X)$ is an at most countable union of intervals of length Δ on which we can reproduce the previous reasoning, we conclude that the solution is identically null. $\square$

It is obvious from the previous proof that we cannot take $\delta = \alpha$ in (2.24). It doesn't mean necessarily that we don't have uniqueness here, it just means that we need a more subtle approach.

Another challenge is given by the following non-uniqueness example. Consider the initial value problem

$$\begin{cases} {}_0D_t^\alpha(x)(t) - \frac{k}{t^\alpha} \cdot x(t) = 0, \, t > 0, \\ \lim_{t \searrow 0}[t^{1-\alpha}x(t)] = 0, \end{cases}$$

where[3] $k = \frac{1}{\Gamma(1-\alpha)} \int_0^1 \frac{\lambda^\alpha}{(1-\lambda)^\alpha} d\lambda$. The problem has two solutions: 0 and t^α.

So, it seems that the "coefficient of $\frac{|x|}{t}$" issue from the case of ordinary differential equations has been replaced for the fractional differential equations by an "exponent of t" issue.

In the spirit of Nagumo's original approach, we shall deal with uniqueness here by prescribing further the behavior of x in the vicinity of 0.

Proposition 2.3. *Set $\beta \in (0, 1 - \alpha)$. Assume that $\delta = \alpha$ in (2.24) and also that*

$$k < \Gamma(\alpha) \left[\int_0^1 \frac{d\lambda}{\lambda^{\alpha+\beta}(1-\lambda)^{1-\alpha}} \right]^{-1}.$$

Then, the only mild solution x of the problem (2.22) such that $\lim_{t \searrow 0}[t^\beta x(t)] = x_1 \in \mathbb{R}$ is the identically null solution[4].

Proof. Suppose, for the sake of contradiction, that x is nontrivial. So, reintroducing $y = y_{-\beta}$ from (2.20) as a continuous function with $y(0) = x_1$ instead of $y(0) = 0$, we deduce that

$$|x(t)| \leq \frac{k}{\Gamma(\alpha)} \int_0^t \frac{|x(s)|}{s^\alpha} \cdot \frac{ds}{(t-s)^{1-\alpha}} \leq \frac{k}{\Gamma(\alpha)} \int_0^t \frac{y(s)}{s^{\alpha+\beta}} \cdot \frac{ds}{(t-s)^{1-\alpha}}$$

---

[3]As before, via (1.3), we have $k = \frac{B(1+\alpha,1-\alpha)}{\Gamma(1-\alpha)} = \frac{1}{\Gamma(1-\alpha)} \cdot \frac{\Gamma(1+\alpha)\cdot\Gamma(1-\alpha)}{\Gamma(2)} = \Gamma(1+\alpha)$.

[4]Notice first that, since $\lim_{t \searrow 0}[t^\beta x(t)]$ exists and is finite, we have $x_0 = \lim_{t \searrow 0}[t^{1-\alpha}x(t)] = \lim_{t \searrow 0}\{t^{1-\alpha-\beta} \cdot [t^\beta x(t)]\} = 0$.

$$\leq y(T) \cdot \frac{k}{\Gamma(\alpha)} \int_0^1 \frac{d\lambda}{\lambda^{\alpha+\beta}(1-\lambda)^{1-\alpha}} \cdot t^{-\beta}, \quad t \in (0, T]. \tag{2.28}$$

We end up with a contradiction

$$y(T) \leq \frac{k}{\Gamma(\alpha)} \int_0^1 \frac{d\lambda}{\lambda^{\alpha+\beta}(1-\lambda)^{1-\alpha}} \cdot y(T) < y(T).$$

Since the number T can be taken arbitrarily large, we conclude that the solution x is trivial. The proof is complete. $\square$

A variant of Proposition 2.3 with improved k is given next.

Proposition 2.4. *Set* $\beta \in (0, 1-\alpha)$. *Assume that* $\delta = \alpha$ *in (2.24) and also that*

$$k = \Gamma(\alpha) \left[\int_0^1 \frac{d\lambda}{\lambda^{\alpha+\beta}(1-\lambda)^{1-\alpha}} \right]^{-1}. \tag{2.29}$$

Then the only mild solution x *of the problem (2.22) such that* $\lim_{t \searrow 0}[t^\beta x(t)] = 0$ *is the identically null solution.*

Proof. Now, as in (2.21), the estimate (2.28) reads as

$$|x(t)| < y_{-\beta}(T) \cdot \frac{k}{\Gamma(\alpha)} \int_0^1 \frac{d\lambda}{\lambda^{\alpha+\beta}(1-\lambda)^{1-\alpha}} \cdot t^{-\beta}$$

$$= y_{-\beta}(T)t^{-\beta}, \quad t \in (0, T].$$

This leads to a contradiction since the maximum of $t^\beta|x(t)|$ is attained at $t = \xi_{T,-\beta}$. $\square$

Observe that (2.29) yields

$$k < \Gamma(\alpha) \left[\int_0^1 \frac{d\lambda}{\lambda^\alpha(1-\lambda)^{1-\alpha}} \right]^{-1} = \frac{1}{\Gamma(1-\alpha)}. \tag{2.30}$$

The fractional variant of Constantin's sharp extension from [Constantin (2010)] of the Nagumo uniqueness criterion reads as follows.

Theorem 2.4. *([Baleanu et al. (2011d), Theorem 1]) Set* $\beta \in (0, 1-\alpha)$ *and suppose there is a continuous, increasing function* $\omega : [0, +\infty) \to [0, +\infty)$ *such that* $\omega(0) = 0$ *and*

$$\frac{1}{\Gamma(\alpha)} \int_0^t \frac{\omega(\varepsilon s^{-\beta})}{s^\alpha(t-s)^{1-\alpha}} \, ds \leq \varepsilon t^{-\beta} \quad \text{for all } t > 0, \, \varepsilon > 0. \tag{2.31}$$

Assume also that

$$|f(t, x)| \leq \frac{\omega(|x|)}{t^\alpha} \quad \text{for all } t > 0, \, x \in \mathbb{R}.$$

Then, the only mild solution x *of the problem (2.22) such that* $\lim_{t \searrow 0}[t^\beta x(t)] = 0$ *is the identically null solution.*

A reformulation of (2.31) reads as

$$\frac{1}{\beta\Gamma(\alpha)} \int_r^{+\infty} \frac{1}{u^{1+\frac{1-\alpha}{\beta}}} \cdot \frac{\omega(u)}{\left[\left(\frac{1}{r}\right)^{\frac{1}{\beta}} - \left(\frac{1}{u}\right)^{\frac{1}{\beta}}\right]^{1-\alpha}} du \le r, \quad r > 0.$$

Taking into account the monotonicity of ω, the inequality (2.31) leads to

$$\frac{\varepsilon}{t^\beta} \ge \frac{\omega\left(\frac{\varepsilon}{t^\beta}\right)}{\Gamma(\alpha)} \int_0^t \frac{ds}{s^\alpha(t-s)^{1-\alpha}} = \Gamma(1-\alpha) \cdot \omega\left(\frac{\varepsilon}{t^\beta}\right),$$

that is $\omega(r) \le \frac{1}{\Gamma(1-\alpha)} \cdot r$ everywhere in $\mathbb{R}_+$. This estimate gives us a hint regarding the sharpness of (2.31) in comparison with (2.24), (2.30).

Proof. As in the proof of Proposition 2.3, if the solution x is supposed to be nontrivial then we obtain

$$|x(t)| \le \frac{1}{\Gamma(\alpha)} \int_0^t \frac{\omega(|x(s)|)}{s^\alpha(t-s)^{1-\alpha}} ds \le \frac{1}{\Gamma(\alpha)} \int_0^t \frac{\omega(y_{-\beta}(s)s^{-\beta})}{s^\alpha(t-s)^{1-\alpha}} ds$$

$$< \frac{1}{\Gamma(\alpha)} \int_0^t \frac{\omega(y_{-\beta}(T)s^{-\beta})}{s^\alpha(t-s)^{1-\alpha}} ds \le y_{-\beta}(T)t^{-\beta}, \quad t \in (0, T].$$

Given the strict inequality in the previous series of inequalities, we get to a contradiction by taking $t = \xi_{T,-\beta}$. Since T is arbitrary, we conclude that the only solution of the problem is the trivial solution. $\square$

2.3   A Wintner-type Existence Interval for the Equation (2.2)

The Peano existence theorem [Peano (1890)] states that *any Cauchy problem attached to a first order ordinary differential equation*, which is defined in a finite dimensional Euclidean space and has a continuous nonlinearity, *possesses at least one (local) continuously differentiable solution*. An *estimate* for the (local) existence interval has been given in the theorem and the proof of this foundational result can be found in almost every general treatise devoted to differential equations, e.g., [Hartman (1964), p. 10]. In 1935, Wintner [Wintner (1935)] has improved this estimate and showed that in some circumstances his variant of the existence interval is optimal. Using a topological device, in 1988, Lee and O'Regan [Lee and O'Regan (1998)] investigated the initial value problem

$$\begin{cases} x'(t) = f(t, x(t)), \, t \ge 0, \\ x(0) = x_0 \in \mathbb{R}^n, \quad n \ge 1, \end{cases} \tag{2.32}$$

under the assumption that $f : \mathbb{R}_+ \times \mathbb{R}^n \to \mathbb{R}^n$ is a continuous function satisfying the restriction

$$\|f(t,x)\| \leq \psi(\|x\|), \quad t \geq 0, \ x \in \mathbb{R},$$

where the (comparison) function $\psi : \mathbb{R}_+ \to (0, +\infty)$ is continuous. They have demonstrated that the problem (2.32) has a solution x in $[0, T]$, where

$$T < T_\infty = \int_{\|x_0\|}^{+\infty} \frac{du}{\omega(u)} \leq +\infty.$$

A slightly different approach has been pursued in [Mustafa (2005); Mustafa and Rogovchenko (2007)] to establish a Wintner-type formulation of the existence interval for the Cauchy problem attached to the differential equation in (2.32).

Our intention in this section is to produce of fractional variant of [Mustafa and Rogovchenko (2007), Theorem 3] in the case of the integral equation (2.2). The fractional counterpart for the *classical existence interval* in Peano's theorem has been developed in [Lakshmikantham and Vatsala (2008), Theorem 3.1], see also the authoritative monograph [Lakshmikantham et al. (2009), Theorem 2.5.1, p. 34], and, for the particular case of f being Lipschitzian with respect to the second variable, in [Diethelm and Ford (2002)], see as well [Kilbas et al. (2006), Theorem 3.26 (iii), p. 206]. Before moving on, let us make a remark on the procedure adopted by Lakshmikantham and Vatsala in [Lakshmikantham and Vatsala (2008), Theorem 3.1], [Lakshmikantham et al. (2009), Theorem 2.4.1, p. 30] for establishing the existence of mild solutions to (2.1). They have employed a less used *memory-like* technique — due to Tonelli [Hartman (1964), p. 23] in the case of ordinary differential equations — to make a uniform approximation of the mild solution. It turns out that such a device comes naturally for the case of "fractionals" as in some applications researchers have used models based on fractional differential equations just for introducing memory (of stress and strain) in the physical processes, see [Metzler et al. (1995), p. 7180].

The existence interval presented in these citations is $[0, T^\star]$, where

$$T^\star = \min\left\{ T, \left(\frac{\alpha b}{M} \right)^{\frac{1}{\alpha}} \right\}, \quad M = \frac{1}{\Gamma(\alpha)} \cdot \sup_{(t,x) \in D} |f(t,x)|.$$

Here, $D = [0, T] \times [x_0 - b, x_0 + b]$ for some $b > 0$.

The reason for the second term in the estimate of $T^\star$ is a consequence of asking that $x(t)$ *from (2.2) be in* $[x_0 - b, x_0 + b]$ *for any* $t \geq 0$. Precisely,

$$|x(t) - x_0| \leq \frac{1}{\Gamma(\alpha)} \cdot \int_0^t \frac{|f(s, x(s))|}{(t-s)^{1-\alpha}} ds$$

$$= \frac{1}{\Gamma(\alpha)} \int_0^t |f(t-s, x(t-s)| \frac{ds}{s^{1-\alpha}} \leq M \cdot \frac{t^\alpha}{\alpha} \leq b$$

which implies that *the mild solution of (2.1) can last for at most*

$$T^\star \leq \left(\frac{\alpha b}{M} \right)^{\frac{1}{\alpha}} \tag{2.33}$$

"units" of t. A Wintner-type estimate of the existence interval enlarges significantly in some circumstances the right-hand part of (2.33).

Let us set several quantities first. We introduce $p, q > 1$ such that $(1-\alpha)q < 1$ and $\frac{1}{p} + \frac{1}{q} = 1$. We also define

$$c = \frac{1}{1-(1-\alpha)q} = \frac{1}{q} \cdot \frac{1}{\frac{1}{q} - 1 + \alpha} = \frac{1}{q} \cdot \frac{1}{\alpha - \frac{1}{p}} = \frac{p}{q} \cdot \frac{1}{\alpha p - 1}$$

$$= \frac{p-1}{\alpha p - 1}, \quad C = \frac{c^{p-1}}{\Gamma(\alpha)},$$

and

$$\lambda = [1-(1-\alpha)q]\frac{p}{q} = \frac{1-(1-\alpha)q}{q-1}$$

$$= \frac{\frac{1}{q} - 1 + \alpha}{1 - \frac{1}{q}} = \frac{\alpha - \frac{1}{p}}{\frac{1}{p}}$$

$$= \alpha p - 1.$$

Consider next the continuous functions $g : [0, T] \times [-b, b] \to \mathbb{R}$ and $w : [0, b] \to [0, +\infty)$ given by

$$g(t, y) = f(t, x_0 + y), \quad w(r) = \sup\{|g(s, y)| : s \in [0, T], |y| \leq r\}.$$

Notice that $w(0) - 0$ *if and only if $f(\cdot, x_0)$ is identically null in $[0, T]$*. By assuming that $w(0) > 0$, we introduce the continuous function $W : [0, b^p] \to [0, +\infty)$ via the formula

$$W(r) = \int_0^r \frac{dv}{\left[w\left(v^{\frac{1}{p}}\right) \right]^p} = p \int_0^{r^{\frac{1}{p}}} \left[\frac{\xi}{w(\xi)} \right]^p \frac{d\xi}{\xi}.$$

The proof of our estimate relies on the fixed point result known as the *Leray-Schauder alternative*, see [Dugundji and Granas (1982), Theorem 5.3, pp. 61–62].

Theorem 2.5. (Leray-Schauder alternative) *Let $O : N \to N$ be a completely continuous (compact) operator acting on the normed linear space N. Then, either there exists $y \in N$ such that*

$$y = O(y)$$

or the set

$$E(O) = \{y \in N : y = \eta \cdot T(y) \text{ for a certain } \eta \in (0,1)\}$$

is unbounded.

We shall take $N = C([0, T^\star], \mathbb{R})$ with the usual *sup*–norm $\|\cdot\|_\infty$, where — recall that $w(0) > 0$ —

$$T^\star = \min\left\{T, \left[\frac{W(b^p)}{C}\right]^{\frac{1}{1+\lambda}}\right\}$$

$$= \min\left\{T, \left[\Gamma(\alpha)\left(\frac{\alpha p - 1}{p - 1}\right)^{p-1} \cdot p \int_0^b \frac{\xi^{p-1}}{[w(\xi)]^p}d\xi\right]^{\frac{1}{\alpha p}}\right\}, \quad (2.34)$$

and

$$O(y)(t) = \frac{1}{\Gamma(\alpha)} \int_0^t \frac{g(s, y(s))}{(t-s)^{1-\alpha}}ds, \quad t \in [0, T^\star], \, y \in N.$$

The compactness of operator O is standard, see [Delbosco and Rodino (1996)] — where Schauder's fixed point theorem has been employed to prove the local existence of solution for (2.2) — and [Kilbas et al. (2006), p. 139].

In fact, let $B \subset N$ be a bounded family of continuous functions:

$$\|b\|_\infty = \sup_{t \in [0,T^\star]} |b(t)| \leq C(B) < +\infty, \quad \text{where } b \in B.$$

Thus,

$$|O(b)(t)| = \left|Q_{1-\alpha, \frac{1}{\Gamma(\alpha)}}g(t, b)\right| \quad (2.35)$$

$$\leq \frac{1}{\Gamma(\alpha)} \int_0^t \frac{ds}{(t-s)^{1-\alpha}} \cdot \sup_{(s,u) \in [0,T^\star] \times [-C(B), C(B)]} |g(s, u)|$$

$$\leq \frac{(T^\star)^\alpha}{\Gamma(1+\alpha)} \cdot \sup_{(s,u) \in [0,T^\star] \times [-C(B), C(B)]} |g(s, u)| < +\infty,$$

where the mapping $t \mapsto g(t, b(t))$ has been denoted with $g(t, b)$ and the quantity $Q_{\beta, A}$ has been defined in (1.30). Consequently, *the set $Q(B)$ is totally (uniformly) bounded in N.*

Further, given $T^\star \geq t_2 > t_1 \geq 0$ and taking into account (2.35), we have

$$|O(b)(t_2) - O(b)(t_1)| \leq D \cdot \frac{1}{\Gamma(\alpha)} \sup_{(s,u) \in [0,T^\star] \times [-C(B), C(B)]} |g(s, u)|$$

$$\times (t_2 - t_1)^{\frac{1}{\max\{q, w\}}},$$

recall (1.45), (1.50), (1.51) for $\xi = \gamma = 0$. In the latter formula, the numbers q and w were introduced at page 15. This means that *the set $O(B)$ is equicontinuous.* According to [Hartman (1964), Selection Theorem 2.3, p. 4], the set $O(B)$ is relatively compact in N. This concludes the proof of compactness as regards the operator O.

We can now state and prove our result.

Theorem 2.6. ([Baleanu and Mustafa (2011), Theorem 2.2]) *Suppose that $f : D \to \mathbb{R}$ is continuous. Then, the following alternative is valid: either the integral equation (2.2) has the constant solution $x = x_0$ throughout $[0, T]$ or it has at least one (non-constant) continuous solution $x(t)$ in $[0, T^\star]$, where $T^\star$ is given by (2.34).*

Proof. Assume that $w(0) > 0$. According to Theorem 2.5, it is enough to establish that the set $E(O)$ is bounded.

Take $y \in E(O)$. Then, we have the estimates

$$|y(t)| \leq \frac{1}{\Gamma(\alpha)} \int_0^t \frac{|g(s, y(s))|}{(t-s)^{1-a}} ds \leq \frac{1}{\Gamma(\alpha)} \int_0^t \frac{w(|y(s)|)}{(t-s)^{1-a}} ds$$

$$\leq \frac{1}{\Gamma(\alpha)} \left\{ \int_0^t \left[\frac{1}{(t-s)^{1-a}} \right]^q ds \right\}^{\frac{1}{q}} \cdot \left\{ \int_0^t [w(|y(s)|)]^p ds \right\}^{\frac{1}{p}}$$

$$= \frac{1}{\Gamma(\alpha)} \left[t^{1-(1-a)q} c \right]^{\frac{1}{q}} \cdot \left\{ \int_0^t [w(|y(s)|)]^p ds \right\}^{\frac{1}{p}}$$

and

$$|y(t)|^p \leq \frac{c^{p-1} t^\lambda}{\Gamma(\alpha)} \cdot \int_0^t [w(|y(s)|)]^p ds$$

$$= C t^\lambda z(t), \quad z(t) = \int_0^t [w(|y(s)|)]^p ds \qquad (2.36)$$

for any $t \in [0, T^\star]$.

Fix now $t_0 \in (0, T^\star]$. We deduce that

$$z'(t) = [w(|y(t)|)]^p \leq \left\{ w\left([C t_0^\lambda z(t)]^{\frac{1}{p}} \right) \right\}^p, \quad t \in [0, t_0],$$

and

$$\frac{[\alpha z(t)]'}{\left\{ w\left([\alpha z(t)]^{\frac{1}{p}} \right) \right\}^p} \leq \alpha, \quad \alpha = C t_0^\lambda.$$

Integrating in $[0, t_0]$, we obtain

$$W(\alpha z(t)) \leq \alpha t \qquad (2.37)$$

and also, by taking $t = t_0$ in (2.37),

$$W(Ct^\lambda z(t)) \leq Ct^{1+\lambda} \leq C(T^\star)^{1+\lambda} \leq W(b^p) \qquad (2.38)$$

for any $t \in [0, T^\star]$.

Since the function $W : [0, b^p] \rightarrow [0, W(b^p)]$ is bijective, by combining the estimates (2.36), (2.38), we conclude that

$$\|y\|_\infty = \sup_{s \in [0, T^\star]} |y(s)| \leq b, \quad y \in E(O).$$

We have established the boundedness of $E(O)$. $\square$

To conclude the section, let us compare the quantities displayed in (2.33), (2.34). Set $x_0, L > 0$ and take

$$f(t, x) = Lx, \quad (t, x) \in D.$$

Thus, $M = \frac{L}{\Gamma(\alpha)}(x_0 + b)$ and $w(r) = L(x_0 + r)$ for any $r \in [0, b]$.

We have thus to compare the quantities

$$H_1(b) = \left[\frac{\Gamma(1+\alpha)b}{L(x_0 + b)} \right]^{\frac{1}{\alpha}}$$

and

$$H_2(b) = \left[\Gamma(\alpha) \left(\frac{\alpha p - 1}{p - 1} \right)^{p-1} \cdot \frac{p}{L^p} \int_0^b \frac{\xi^{p-1}}{(x_0 + \xi)^p} d\xi \right]^{\frac{1}{\alpha p}},$$

where $b \geq 0$.

Introduce $Q_i(b) = [H_i(b)]^{\alpha p}$ for $b \geq 0$, where $i = 1, 2$. Then, we notice that

$$\lim_{b \to +\infty} Q_1(b) = \left[\frac{\Gamma(1+\alpha)}{L} \right]^p. \qquad (2.39)$$

On the other hand, since for $b \geq x_0$ it is clear that

$$\int_{x_0}^b \frac{\xi^{p-1}}{(x_0 + \xi)^p} d\xi \geq \left(\frac{1}{2} \right)^p \int_{x_0}^b \frac{d\xi}{\xi} = \left(\frac{1}{2} \right)^p \cdot \log \left(\frac{b}{x_0} \right),$$

we get

$$\lim_{b \to +\infty} Q_2(b) = +\infty. \qquad (2.40)$$

The formulas (2.39), (2.40) show that *the estimate (2.34) is much better than its classical counterpart (2.33) for large values of b,* similarly to the case of ordinary differential equations.

Chapter 3

Position of the Zeros, the Bihari Inequality, and the Asymptotic Behavior of Solutions for the Differential Equations of Order α

Several unrelated results, presented in the following, enrich the beautiful panorama of fractional differential equations of *order less than* 1 that has been given a modest sketch in the previous chapter.

3.1 A Fite-type Length Criterion for Fractional Disconjugacy

Given the real numbers $a < b$, consider the ordinary differential system $(m \geq 2)$

$$\begin{cases} x'_1(t) = f_1(t, x_1(t), \cdots, x_m(t)) \\ \qquad \cdots\cdots\cdots\cdots\cdots \\ x'_m(t) = f_m(t, x_1(t), \cdots, x_m(t)). \end{cases} \tag{3.1}$$

The functions $f_i : [a, b] \times \mathbb{R}^m \to \mathbb{R}$ are Lipschitzian, that is

$$|f_i(t, y_1, \ldots, y_m) - f_i(t, x_1, \ldots, x_m)| \leq \sum_{k=1}^{m} L_k |y_k - x_k|, \quad \text{where } t \in [a, b]$$

and x_i, $y_i \in \mathbb{R}$ for all $1 \leq i \leq m$. Here, $\sum_{k=1}^{m} L_k > 0$ and all the L_k's are nonnegative.

Theorem 3.1. (W. B. Fite, [Fite (1918), Theorem VII]) *Assume that* $(x_1, \ldots, x_m)$ *and* $(y_1, \ldots, y_m)$ *are distinct (classical) solutions of the system (3.1). If there exist the points* $(t_i)_{i \in \overline{1,m}}$ *in* (a, b) *such that* $x_i(t_i) = y_i(t_i)$, *where $1 \leq i \leq m$, then*

$$b - a \geq \frac{1}{\sum\limits_{k=1}^{m} L_k}. \tag{3.2}$$

Proof. Set $\varepsilon > 0$. Observe that

$$|y_i(t) - x_i(t)|$$

$$= |(y_i - x_i)(t) - (y_i - x_i)(t_i)| = \left| \int_{t_i}^{t} (y_i - x_i)'(s) ds \right|$$

$$\leq \left| \int_{t_i}^{t} |f_i(s, y_1(s), \ldots, y_m(s)) - f_i(s, x_1(s), \ldots, x_m(s))| ds \right|$$

$$\leq \sum_{k=1}^{m} (L_k + \varepsilon) \cdot \sup_{\tau \in [\min\{t, t_i\}, \max\{t, t_i\}]} |y_k(\tau) - x_k(\tau)| \cdot \left| \int_{t_i}^{t} ds \right|$$

$$\leq \left[\sum_{k=1}^{m} (L_k + \varepsilon) M_k \right] \cdot (b - a), \quad t \in [a, b],$$

where $M_k = \sup\limits_{\tau \in [a,b]} |y_k(\tau) - x_k(\tau)|$. The quantity ε has been introduced to compensate the sum for the (eventual) null terms L_k.

Further,

$$M_i \leq \sum_{k=1}^{m} (L_k + \varepsilon) M_k \cdot (b - a), \quad 1 \leq i \leq m.$$

By summation, we get

$$\sum_{i=1}^{m} (L_i + \varepsilon) M_i \leq \left[\sum_{k=1}^{m} (L_k + \varepsilon) M_k \right] \left(\sum_{i=1}^{m} L_i + m \cdot \varepsilon \right) (b - a).$$

We have either $\sum\limits_{k=1}^{m} (L_k + \varepsilon) M_k = 0$, which means the solutions coincide everywhere in $[a, b]$, or the estimate

$$\frac{1}{\sum\limits_{k=1}^{m} L_k + m\varepsilon} \leq b - a$$

holds true. In the latter case, by making $\varepsilon \searrow 0$, we get to (3.2). $\square$

In the particular circumstances when $f_i(t, 0, \cdots, 0) = 0$ for any $t \in [a, b]$ and any $i \in \overline{1, m}$, Theorem 3.1 provides a lower bound for the *length of the interval* where each of the components x_i of some solution $x = (x_1, \ldots, x_m)$ to (3.1) can have at least one zero. This type of investigation in the case of linear ordinary differential equations of order m has been inaugurated by Fite in 1917 [Fite (1918), p. 350]. When $m = 2$, the Fite length criterion provides an estimate of the *disconjugacy interval* of some ordinary differential equations, see [Coppel (1971), Chapter 1].

To state a fractional variant of the Fite length criterion, set the real numbers $a < b < c$ and consider the system of fractional differential equations

$$\begin{cases} {}_aD_t^\alpha(f)(t) = G(t)g(t) + Q(t) \\ {}_aD_t^\alpha(g)(t) = R(t)f(t) + V(t), \end{cases} \quad t \in [b, c], \qquad (3.3)$$

where the coefficients $G, Q, R, V : [a, +\infty) \to \mathbb{R}$ are continuous and G, R are bounded: $\|G\|_{L^\infty([a,+\infty),\mathbb{R})} + \|R\|_{L^\infty([a,+\infty),\mathbb{R})} < +\infty$.

As before, we introduce the integral counterpart of the system (3.3), namely

$$\begin{cases} f(t) = \frac{f_a}{(t-a)^{1-\alpha}} + \frac{1}{\Gamma(\alpha)} \int_a^t \frac{G(s)g(s)+Q(s)}{(t-s)^{1-\alpha}} ds \\ g(t) = \frac{g_a}{(x-a)^{1-\alpha}} + \frac{1}{\Gamma(\alpha)} \int_a^t \frac{R(s)f(s)+V(s)}{(t-s)^{1-\alpha}} ds, \end{cases} \quad t \in (a, c], \qquad (3.4)$$

where $f_a, g_a \in \mathbb{R}$. Any solution $(f, g) \in \left(\mathcal{RL}^{1-\alpha}\right)^2$ of the integral system (3.4) will be referred to in this section as a *mild solution* to (3.3). Recall (1.31) for $\gamma = 1 - \alpha$.

Theorem 3.2. ([Abdeljawad et al. (2010), Theorem 1]) *Set* $\alpha \in \left(\frac{1}{2}, 1\right)$, $p > 1$ *with* $(1-\alpha)p < \frac{1}{2}$ *and take* $q = \frac{p}{p-1}$. *Assume that there exist the mild solutions* $(f_i, g_i)_{i \in \overline{1,2}}$ *of the system (3.3) such that*

$$f_1(t_1) = f_2(t_1), \quad g_1(t_2) = g_2(t_2)$$

for some $t_1, t_2 \in [b, c]$.
   Then,

$$m \cdot (c-a)^\alpha \frac{\max\left\{(c-a)^{\frac{1}{q}}, (c-a)^{1-\alpha}\right\}}{\min\{(c-a)^{\frac{1}{q}}, (c-a)^{1-\alpha}\}} \geq \frac{\Gamma(\alpha)}{2^{2(2-\alpha)} + B(\alpha, \alpha)},$$

where $m = \max\{\|G\|_{L^\infty([a,+\infty),\mathbb{R})}, \|R\|_{L^\infty([a,+\infty),\mathbb{R})}\}$.

To use the estimates from page 15, introduce the quantities

$$0 < \gamma = \beta = 1 - \alpha < \frac{1}{2}.$$

There exists $p = v > 1$ such that

$$\gamma p = \beta v < \frac{1}{2}.$$

Also, $q = w$.

An important consequence of this choice of constants is that

$$1 - \beta - \gamma - \frac{1}{\max\{q, w\}} = 1 - 2\gamma - \frac{1}{q} = \frac{1}{p} - 2\gamma > 0,$$

leading to — recall (1.51) — a quantity D which is independent of $b \in (a, c)$.

Also, since $c(p, \gamma, \gamma) = \frac{2^{2\gamma - \frac{1}{p}}}{(1-\gamma p)^{\frac{1}{p}}} < \frac{2^{2\gamma - \frac{1}{p}}}{\left(\frac{1}{2}\right)^{\frac{1}{p}}} = 2^{2\gamma}$, where c has been defined in (1.38), we have (remember (1.39))

$$C = 2[c(p, \beta, \gamma) + c(v, \gamma, \beta)] = 4c(p, \gamma, \gamma) < 2^{2(1+\gamma)} = 2^{2(2-\alpha)}$$

and respectively — set $\xi = \beta$ —

$$\begin{aligned}
D &< 2^{2(2-\alpha)}(c - a)^{\alpha - \frac{1}{q}} + B(\alpha, \alpha)(c - a)^{2\alpha - 1} \\
&\leq [2^{2(2-\alpha)} + B(\alpha, \alpha)] \cdot \frac{(c - a)^{\alpha}}{\min\{(c - a)^{\frac{1}{q}}, (c - a)^{1-\alpha}\}}.
\end{aligned} \tag{3.5}$$

Proof. Take $b \leq t \leq t_1 \leq c$. By means of (1.50) and reminding of the notation (1.32), we deduce that

$$\begin{aligned}
&|(t - a)^{1-\alpha}(f_2 - f_1)(t)| \\
&= |(t_1 - a)^{1-\alpha}(f_2 - f_1)(t_1) - (t - a)^{1-\alpha}(f_2 - f_1)(t)| \\
&= \left| \frac{1}{\Gamma(\alpha)} \cdot [(t - a)^{1-\alpha}(Q_{1-\alpha,G}(g_2 - g_1))(t) \right. \\
&\quad \left. - (t_1 - a)^{1-\alpha}(Q_{1-\alpha,G}(g_2 - g_1))(t_1)] \right| \\
&\leq \frac{D}{\Gamma(\alpha)} \cdot \|G\|_{L^\infty([a,c],\mathbb{R})} \|g_2 - g_1\|_{L^\infty(a,1-\alpha;c)} \\
&\quad \times \max\left\{ (t_1 - t)^{\frac{1}{q}}, (t_1 - t)^{1-\alpha} \right\} \\
&\leq \frac{D\|G\|_{L^\infty([a,+\infty),\mathbb{R})}}{\Gamma(\alpha)} \max\left\{ (c - b)^{\frac{1}{q}}, (c - b)^{1-\alpha} \right\} \cdot \|g_2 - g_1\|_{L^\infty(a,1-\alpha;c)} \\
&\leq \frac{D\|G\|_{L^\infty([a,+\infty),\mathbb{R})}}{\Gamma(\alpha)} \max\left\{ (c - a)^{\frac{1}{q}}, (c - a)^{1-\alpha} \right\} \cdot \|g_2 - g_1\|_{L^\infty(a,1-\alpha;c)}.
\end{aligned}$$

The case $b \leq t_1 \leq t \leq c$ leads to the same conclusion. So,

$$\begin{aligned}
&\|f_2 - f_1\|_{L^\infty(a,1-\alpha;b,c)} \\
&\leq E \cdot \|G\|_{L^\infty([a,+\infty),\mathbb{R})} \cdot \|g_2 - g_1\|_{L^\infty(a,1-\alpha;c)} \\
&\leq E \cdot \max\{\|G\|_{L^\infty([a,+\infty),\mathbb{R})}, \|R\|_{L^\infty([a,+\infty),\mathbb{R})}\} \\
&\quad \times \|g_2 - g_1\|_{L^\infty(a,1-\alpha;c)},
\end{aligned} \tag{3.6}$$

where, via (3.5),

$$E = \frac{D}{\Gamma(\alpha)} \max\left\{(c-a)^{\frac{1}{q}}, (c-a)^{1-\alpha}\right\}$$

$$\leq \frac{2^{2(2-\alpha)} + B(\alpha,\alpha)}{\Gamma(\alpha)} \cdot (c-a)^{\alpha} \frac{\max\left\{(c-a)^{\frac{1}{q}}, (c-a)^{1-\alpha}\right\}}{\min\{(c-a)^{\frac{1}{q}}, (c-a)^{1-\alpha}\}}. \quad (3.7)$$

Similarly,

$$\|g_2 - g_1\|_{L^\infty(a,1-\alpha;b,c)}$$
$$\leq E \cdot \|R\|_{L^\infty([a,+\infty),\mathbb{R})} \cdot \|f_2 - f_1\|_{L^\infty(a,1-\alpha;c)}$$
$$\leq E \cdot \max\{\|G\|_{L^\infty([a,+\infty),\mathbb{R})}, \|R\|_{L^\infty([a,+\infty),\mathbb{R})}\}$$
$$\times \|f_2 - f_1\|_{L^\infty(a,1-\alpha;c)}. \quad (3.8)$$

In conclusion, as E is independent of b, by taking into account (3.6), (3.8) and (1.33), we get

$$X \leq E \cdot \max\{\|G\|_{L^\infty([a,+\infty),\mathbb{R})}, \|R\|_{L^\infty([a,+\infty),\mathbb{R})}\} \cdot X,$$

where $X = \max\{\|f_2 - f_1\|_{L^\infty(a,1-\alpha;c)}, \|g_2 - g_1\|_{L^\infty(a,1-\alpha;c)}\}$.

The proof is complete. $\square$

The conclusion of Theorem 3.1 can be adapted easily to the sequential[1] fractional differential equation

$$_aD_t^\alpha(_aD_t^\alpha f)(t) + P \cdot f(t) = 0, \quad t \in [b,c],$$

where $\alpha \in \left(\frac{1}{2}, 1\right)$, $P > 0$ and the solution f is given by the restrictions f, $_aD_t^\alpha f \in \mathcal{RL}^{1-\alpha}$. The result provides us with a lower bound for the distance between a and c, where the latter is such that both (nontrivial) functions f and $_aD_t^\alpha f$ have at least one zero in (a,c).

3.2 The Bihari Inequality

To get full understanding of the asymptotic integration methods for the fractional differential equations of order $1 + \alpha$, we need to detail first an ingenious asymptotic technique that has proved to be salutary in the case of the ordinary differential equations of second order. In the latter situation, the general integral inequality due to Bihari[2] is of utmost importance.

---

[1]See [Podlubny (1999a), p. 87].

[2]Besides its own interest, the Bihari (Gronwall-like) inequality has been employed for dealing with technical problems in the field of fractional differential equations by various authors, e.g., [Furati and Tatar (2005), p. 1031].

We start with the *direct inequality*. Assume that $u, a : [t_0, T) \to [0, +\infty)$, where $t_0 \geq 1$, $T \leq +\infty$, and $g : [0, +\infty) \to [0, +\infty)$ are continuous functions, g being also non-decreasing and such that $g(x) > 0$ for all $x \geq k \geq 0$.

We claim that *if there exist $c \geq k$ such that*

$$u(t) \leq c + \int_{t_0}^t a(s)g(u(s))ds, \quad t \in [t_0, T), \tag{3.9}$$

and respectively $c < C \leq +\infty$ such that

$$\int_{t_0}^T a(s)ds < \int_c^C \frac{du}{g(u)}, \tag{3.10}$$

then

$$u(t) \leq G^{-1}\left(G(c) + \int_{t_0}^t a(s)ds\right) < C, \quad t \in [t_0, T), \tag{3.11}$$

where

$$G(x) = \int_k^x \frac{du}{g(u)}, \quad x \geq k. \tag{3.12}$$

To prove this claim, introduce the quantity — nothing but the right-hand member of (3.9) —

$$y(t) = c + \int_{t_0}^t a(s)g(u(s))ds, \quad t \in [t_0, T),$$

and recast (3.9) as

$$u(t) \leq y(t), \quad t \in [t_0, T). \tag{3.13}$$

Notice that $y(t_0) = c$ and $y(t) \geq c \geq k$. Consequently, $g(y(t)) > 0$ in $[t_0, T)$.

We have

$$y'(t) = a(t)g(u(t)) \leq a(t)g(y(t))$$

and

$$\frac{y'(t)}{g(y(t))} \leq a(t), \quad t \in [t_0, T).$$

By integration, we deduce that

$$G(y(t)) - G(c) = \int_c^{y(t)} \frac{du}{g(u)} = \int_{t_0}^t \frac{y'(s)}{g(y(s))}ds \tag{3.14}$$

$$\leq \int_{t_0}^t a(s)ds. \tag{3.15}$$

Observe now that — recall (3.10) —

$$G(c) + \int_{t_0}^{t} a(s)ds < G(c) + \int_{c}^{C} \frac{du}{g(u)} = G(C), \quad t \in [t_0, T). \quad (3.16)$$

The function $G : [k, C] \to [0, G(C)]$ is bijective and increasing. We tacitly assume that C is bounded. The case when $C = +\infty$ can be handled almost verbatim.

Further, the function y being continuous and non-decreasing, and $y(t_0) = c \in [k, C)$, it is obvious that $y(t) \in [k, C)$ on a certain interval I to the right of t_0. By taking into account (3.15), (3.16), we deduce that $[t_0, T) \subseteq I$. Moreover,

$$y(t) \leq G^{-1}\left(G(c) + \int_{t_0}^{t} a(s)ds\right) < C, \quad t \in [t_0, T).$$

This follows from the monotonicity of G^{-1}.

The estimate (3.11) is a consequence of (3.13). The validity of our claim is now established.

The *Bihari inequality* (3.9), (3.11) was presented first in [Bihari (1956), pp. 83–85]. The case of reverse sign in (3.9) is discussed in [Bihari (1956), pp. 92–93]. The cases when $C = +\infty$ and the right-hand member of (3.10) is divergent are discussed in [Bihari (1957), p. 262].

A similar inequality, namely

$$u(t) \leq b(t)\left[c + \int_{t_0}^{t} a(s)g(u(s))ds\right], \quad t \in [t_0, T),$$

has been studied by Golomb [Golomb (1958), p. 273, Eq. (2.6)] when the continuous nonlinearity g is assumed to be non-decreasing, with $g(x) > 0$ for every $x > 0$, and also sub-multiplicative. The latter property yields the key estimate regarding G given below — here, $c > 0$ and $k = 1$ —

$$G(u + v) - G(u) \geq \frac{u}{g(u)} \cdot G\left(1 + \frac{v}{u}\right), \quad u > 0, \ v \geq 0,$$

see [Golomb (1958), p. 274]. This inequality is used in conjunction with (3.14) to prove that

$$u(t) \leq b(t) \cdot c\, G^{-1}\left(\frac{g(c)}{c} \cdot \int_{t_0}^{t} a(s)g(b(s))ds\right), \quad t \in [t_0, T).$$

The case of a sub-additive nonlinearity g is undertaken in [Muldowney and Wong (1968), p. 491]. For other developments, see [Dannan (1985); Pinto (1998); Rogovchenko (1998)].

We shall end the discussion regarding the direct Bihari inequality with a remark concerning the middle term in the double inequality (3.11). For any $\alpha \in [0, C - k)$, introduce the quantity

$$G_\alpha(x) = \int_{k+\alpha}^x \frac{du}{g(u)}, \quad x \in [k, C).$$

Obviously, the function $G_\alpha : [k, C) \to \left[-\int_k^{k+\alpha} \frac{du}{g(u)}, \int_{k+\alpha}^C \frac{du}{g(u)} \right)$ is bijective and increasing.

We notice also that, given $\alpha, \beta \in [0, C - k)$,

$$G_\beta(x) = G_\alpha(x) + q, \quad q = \int_{k+\beta}^{k+\alpha} \frac{du}{g(u)}, \tag{3.17}$$

and respectively

$$x = G_\beta^{-1}(G_\beta(x)) = G_\beta^{-1}(G_\alpha(x) + q) = G_\alpha^{-1}(G_\alpha(x)), \quad x \in [k, C).$$

The latter identity reads as

$$G_\beta^{-1}(y + q) = G_\alpha^{-1}(y), \quad y \in G_\alpha([k, C)). \tag{3.18}$$

Now, by means of (3.17), (3.18), we get — see also [Bihari (1956), p. 84] —

$$G_\alpha^{-1}\left(G_\alpha(c) + \int_{t_0}^t a(s)ds \right) = G_\alpha^{-1}\left(G_\beta(c) - q + \int_{t_0}^t a(s)ds \right)$$

$$= G_\beta^{-1}\left(\left[G_\beta(c) - q + \int_{t_0}^t a(s)ds \right] + q \right)$$

$$= G_\beta^{-1}\left(G_\beta(c) + \int_{t_0}^t a(s)ds \right). \tag{3.19}$$

The remark based on the previous computations is that, *given the relations (3.9), (3.12), there is no need to ask that $c \geq k$, the only restriction being given by $c, k > \sup\{x \geq 0 : g(x) = 0\}$*.

The dual inequality. Let the function $g : [0, +\infty) \to [0, +\infty)$ be continuous, bijective and increasing (of course, these conditions yield $g(x) > 0$ for all $x > 0$). Assume also that

$$\int_0^1 \frac{du}{g^{-1}(u)} < +\infty$$

and introduce $H : [0, +\infty) \to [0, +\infty)$ given by the formulas

$$H(0) = 0 \quad \text{and} \quad H(x) = \int_0^x \frac{du}{g^{-1}(u)}, \quad x > 0.$$

Introduce $0 < c < C \leq +\infty$, $T \leq +\infty$ and the continuous function $a : [t_0, T) \to [0, +\infty)$ such that

$$\int_{t_0}^{T} a(s)ds < \int_{c}^{C} \frac{du}{g(u)}.$$

We claim that, given the continuous function $u : [t_0, T) \to [0, +\infty)$, *(i) if*

$$g(u(t)) \leq c + \int_{t_0}^{t} a(s)u(s)ds, \quad t \in [t_0, T), \tag{3.20}$$

then we have

$$u(t) \leq (H \circ g)^{-1} \left(H(c) + \int_{t_0}^{t} a(s)ds \right), \quad t \in [t_0, T); \tag{3.21}$$

(ii) if

$$g(u(t)) \geq c + \int_{t_0}^{t} a(s)u(s)ds, \quad t \in [t_0, T),$$

then we have

$$u(t) \geq (H \circ g)^{-1} \left(H(c) + \int_{t_0}^{t} a(s)ds \right), \quad t \in [t_0, T).$$

We shall prove only part (i). Then, for $y(t) = \int_{t_0}^{t} a(s)u(s)ds$, with $t \in [t_0, T)$, the inequality (3.20) reads as

$$u(t) \leq g^{-1}(c + y(t)). \tag{3.22}$$

Notice also that $y(t_0) = 0$.

We deduce that

$$\frac{(c+y)'(t)}{g^{-1}(c+y(t))} \leq a(t), \quad t \in [t_0, T),$$

and further, by integration, we get

$$\int_{c}^{c+y(t)} \frac{du}{g^{-1}(u)} \leq \int_{t_0}^{t} a(s)ds.$$

The latter estimate is recast as

$$H(c + y(t)) \leq H(c) + \int_{t_0}^{t} a(s)ds < H(C), \quad t \in [t_0, T).$$

In particular,

$$c + y(t) \leq H^{-1} \left(H(c) + \int_{t_0}^{t} a(s)ds \right). \tag{3.23}$$

It is obvious now that the estimate (3.21) is a consequence of the inequalities (3.22), (3.23).

The following particular case of the *dual Bihari inequality* (3.20), (3.21) is used quite frequently in stability theory and it has been discovered by Ouyang [Ouyang (1957)].

If x, $A : [t_0, T) \to \mathbb{R}$ are continuous functions such that

$$\frac{[x(t)]^2}{2} \leq \frac{[x(t_0)]^2}{2} + \int_{t_0}^t A(s)x(s)ds, \quad t \in [t_0, T),$$

then

$$|x(t)| \leq |x(t_0)| + \int_{t_0}^t |A(s)|ds, \quad t \in [t_0, T).$$

To establish Ouyang's inequality , it is enough to take

$$g(x) = \frac{x^2}{2} = H^{-1}(x), \quad g^{-1}(x) = \sqrt{2x} = H(x)$$

and

$$u(t) = |x(t)|, \quad a(t) = |A(t)|, \quad c = g(u(t_0))$$

in (3.20).

We shall recall in the end of the discussion regarding the dual Bihari inequality certain connections between g and g^{-1}.

Let $g : [a, +\infty) \to [b, +\infty)$ be a bijective, increasing C^1–function. Here, $a, b > 0$ — obviously, $g(a) = b$ —.

Then,

$$\int_a^x \frac{du}{g(u)} = -\frac{a}{g(a)} + \frac{x}{g(x)} + \int_b^{g(x)} \frac{g^{-1}(u)}{u^2}du, \quad x \geq a, \quad (3.24)$$

and

$$\int_b^y \frac{du}{g^{-1}(u)} = -\frac{b}{g^{-1}(b)} + \frac{y}{g^{-1}(y)} + \int_a^{g^{-1}(y)} \frac{g(v)}{v^2}dv, \quad y \geq b. \quad (3.25)$$

The formulas (3.24), (3.25) can be established easily by differentiation with respect to x and respectively y in both sides. Also, the smoothness of g can be reduced to merely continuity if we use the Riemann-Stieltjes integration, see, for instance, [Pucci and Serrin (1991), p. 104].

Going further, let us show that $\int_a^{+\infty} \frac{du}{g(u)} < +\infty$ yields $\int_b^{+\infty} \frac{dv}{g^{-1}(v)} = +\infty$ and $\int_b^{+\infty} \frac{dv}{g^{-1}(v)} < +\infty$ yields $\int_a^{+\infty} \frac{du}{g(u)} = +\infty$.

In fact, if $\int_a^{+\infty} \frac{du}{g(u)} < +\infty$ then we can use the following classical trick, see [Kiguradze and Kvinikadze (1982), p. 72],

$$\frac{x}{g(x)} = \frac{1}{\ln 2} \cdot \frac{x}{g(x)} \cdot \int_{\frac{x}{2}}^{x} \frac{du}{u} = \frac{1}{\ln 2} \cdot \frac{\left(2 \cdot \frac{x}{2}\right)}{g(x)} \cdot \int_{\frac{x}{2}}^{x} \frac{du}{u}$$

$$\leq \frac{2}{\ln 2} \cdot \int_{\frac{x}{2}}^{x} \frac{u}{g(u)} \cdot \frac{du}{u}$$

$$\leq \frac{2}{\ln 2} \cdot \int_{\frac{x}{2}}^{+\infty} \frac{du}{g(u)}. \tag{3.26}$$

Now, since $\lim_{x \to +\infty} \int_{\frac{x}{2}}^{+\infty} \frac{du}{g(u)} = 0$, the estimate (3.26) yields

$$\lim_{x \to +\infty} \frac{x}{g(x)} = 0.$$

This formula can be recast as $\lim_{y \to +\infty} \frac{g^{-1}(y)}{y} = 0$. Consequently,

$$\lim_{y \to +\infty} \frac{y}{g^{-1}(y)} = +\infty. \tag{3.27}$$

Since, by means of (3.26),

$$\frac{y}{g^{-1}(y)} \leq \frac{2}{\ln 2} \cdot \int_{\frac{y}{2}}^{+\infty} \frac{dv}{g^{-1}(v)} \leq +\infty,$$

the estimate (3.27) leads to $\int_b^{+\infty} \frac{dv}{g^{-1}(v)} = +\infty$.

A different proof of (3.27), based on Young's inequality, can be adapted from the computations done in [Pucci and Serrin (1991), p. 104]. The other claim is established similarly.

3.3 Asymptotic Integration of the Differential Equations of Orders 1 and α

Consider the nonlinear ordinary differential equation

$$x'(t) = a(t)w\left(\frac{x(t)}{t}\right), \quad t \geq t_0 \geq 1, \tag{3.28}$$

where $a : [t_0, +\infty) \to (0, +\infty)$, $w : \mathbb{R} \to \mathbb{R}$ are continuous and w satisfies the sign condition

$$u \cdot w(u) > 0 \quad \text{for all } u \neq 0.$$

Furthermore, assume that the restriction of w to $(0, +\infty)$ is non-decreasing. It is obvious that, since 0 is the only zero of w, we must have

$$w(u) > 0 \quad \text{for all } u > 0. \tag{3.29}$$

Given the numbers $p, q > 1$ such that $\frac{1}{p} + \frac{1}{q} = 1$, introduce the quantities

$$b(t) = [a(t)]^p, \quad [d(T)]^{\frac{1}{q}} = \sup_{t \in [t_0, T]} \left(\frac{\int_{t_0}^t b(s)\,ds}{t^p} \right)^{\frac{1}{p}}, \quad T \geq t \geq t_0.$$

Introduce also the quantities

$$c > 0, \quad \mathcal{W}(u) = \int_0^u \frac{d\xi}{\left[w\left(c + \xi^{\frac{1}{q}} \right) \right]^q}, \, u \geq 0$$

and suppose that

$$\lim_{u \to +\infty} \frac{\mathcal{W}(u)}{u^{\alpha}} = +\infty \tag{3.30}$$

for a certain $\alpha \in (0, 1)$.

The connection between a and w is given by the restriction

$$\lim_{T \to +\infty} T \cdot d(T) = +\infty. \tag{3.31}$$

We claim that *all the continuable solutions x with $x(t_0) = x_0 > 0$ and $c = \frac{x_0}{t_0}$ have the asymptotic behavior given by*

$$\lim_{t \to +\infty} \frac{x(t)}{t} \cdot [td(t)]^{-\frac{1}{\alpha q}} = 0. \tag{3.32}$$

Since $x(t_0) > 0$ and $x'(t_0) = a(t_0)w(c) > 0$, it is easy to see that any solution $x(t)$ of equation (3.28) starting from $x(t_0) = x_0$ is positive and increasing throughout $[t_0, +\infty)$.

The integral identity

$$x(t) = x_0 + \int_{t_0}^t a(s)w\left(\frac{x(s)}{s} \right) ds, \quad t \geq t_0,$$

is recast as

$$y(t) = \int_{t_0}^t a(s)w\left(\frac{x_0 + y(s)}{s} \right) ds, \quad \text{where } y(t) = x(t) - x_0, \, t \geq t_0.$$

Notice that $y(t) > 0$ for $t > t_0$.

An application of Hölder's inequality in integral form leads to

$$y(t) \leq \left(\int_{t_0}^t [a(s)]^p\,ds \right)^{\frac{1}{p}} \cdot \left(\int_{t_0}^t \left[w\left(c + \frac{y(s)}{s} \right) \right]^q ds \right)^{\frac{1}{q}}$$

and respectively to

$$\frac{y(t)}{t} \leq \left(\frac{\int_{t_0}^{t} b(s)ds}{t^p}\right)^{\frac{1}{p}} \cdot [z(t)]^{\frac{1}{q}}, \quad z(t) = \int_{t_0}^{t} \left[w\left(c + \frac{y(s)}{s}\right)\right]^q ds.$$

We get

$$\frac{y(t)}{t} \leq [d(T) \cdot z(t)]^{\frac{1}{q}}, \quad t \in [t_0, T], \ T < +\infty. \tag{3.33}$$

We have next that

$$z'(t) = \left[w\left(c + \frac{y(t)}{t}\right)\right]^q \leq \left[w\left(c + [d(T) \cdot z(t)]^{\frac{1}{q}}\right)\right]^q, \quad t \in [t_0, T],$$

by means of (3.33).

Now,

$$\frac{[d(T)z]'(t)}{\left[w\left(c + [d(T)z(t)]^{\frac{1}{q}}\right)\right]^q} \leq d(T).$$

An integration over $[t_0, t]$, where $t \leq T$, yields — notice that $z(t_0) = 0$ and recall (3.31) —

$$\mathcal{W}(d(T)z(t)) - \mathcal{W}(0) = \int_{t_0}^{t} \frac{[d(T)z]'(s)}{\left[w\left(c + [d(T)z(s)]^{\frac{1}{q}}\right)\right]^q} ds$$

$$\leq d(T)(t - t_0) \leq d(T)T$$

for every $t_0 \leq t \leq T < +\infty$.

We deduce further, via (3.33), that

$$\mathcal{W}\left(\left[\frac{y(T)}{T}\right]^q\right) \leq \mathcal{W}(d(T)z(T)) \leq d(T)T,$$

respectively that

$$\frac{\mathcal{W}\left(\left[\frac{y(T)}{T}\right]^q\right)}{\left[\frac{y(T)}{T}\right]^{\alpha q}} \cdot \left\{\frac{y(T)}{T[d(T)T]^{\frac{1}{\alpha q}}}\right\}^{\alpha q} \leq 1, \quad T \geq t_0.$$

If we suppose, for the sake of contradiction, that (3.32) doesn't hold true then there exist $\varepsilon > 0$ and an increasing, unbounded from above, sequence $(T_n)_{n \geq 1}$ taken from $(t_0, +\infty)$ such that

$$\frac{y(T_n)}{T_n[d(T_n)T_n]^{\frac{1}{\alpha q}}} \geq \varepsilon, \quad n \geq 1.$$

This estimate can be recast as

$$\frac{y(T_n)}{T_n} \geq \varepsilon \cdot [d(T_n)T_n]^{\frac{1}{\alpha q}}. \tag{3.34}$$

By taking into account (3.31), (3.34), we obtain that

$$\lim_{n \to +\infty} \frac{y(T_n)}{T_n} = +\infty.$$

In conclusion, since

$$\frac{\mathcal{W}\left(\left[\frac{y(T_n)}{T_n}\right]^q\right)}{\left[\frac{y(T_n)}{T_n}\right]^{\alpha q}} \cdot \varepsilon^{\alpha q} \leq 1, \quad n \geq 1,$$

we arrive at a contradiction — recall (3.30) —.

The hypotheses which lead to the estimate (3.32) being complicate, it is natural to wonder about their efficiency. The following example will show that *the asymptotic behavior described by (3.32) is not raw.*

Set the numbers $\zeta > 0$, $\alpha \in (0,1)$ and $\varepsilon \in \left(0, \frac{1-\alpha}{\alpha}\right)$. Consider the equation

$$x'(t) = t^{\zeta + (1+\varepsilon)\alpha - 1} \cdot [x(t)]^{1 - (1+\varepsilon)\alpha}, \quad t \geq t_0 \geq 1. \tag{3.35}$$

It is obvious that $1 > (1 + \varepsilon)\alpha$. We have also

$$a(t) = t^\zeta, \quad w(u) = u^{1 - (1+\varepsilon)\alpha} \text{ for every } u \geq 0.$$

Fix $p > 1$ such that — if $\zeta \in (0,1)$ — $(1 - \zeta)p \leq 1$ and introduce $q = \frac{p}{p-1}$.

To verify the hypothesis (3.30), introduce $u_c = c^q$. Now,

$$\int_0^u \frac{d\xi}{\left[w\left(c + \xi^{\frac{1}{q}}\right)\right]^q} \geq \int_{u_c}^u \frac{d\xi}{\left[w\left(2 \cdot \xi^{\frac{1}{q}}\right)\right]^q} = 2^{-q[1-(1+\varepsilon)\alpha]} \cdot \int_{u_c}^u \frac{d\xi}{\xi^{1-(1+\varepsilon)\alpha}}$$

$$= \frac{2^{-q[1-(1+\varepsilon)\alpha]}}{(1+\varepsilon)\alpha} \cdot \left(u^{(1+\varepsilon)\alpha} - u_c^{(1+\varepsilon)\alpha}\right) = +\infty \cdot u^\alpha \quad \text{when } u \to +\infty.$$

To check the hypothesis (3.31), let us start by observing that

$$\frac{d}{dt}\left[\frac{\int_{t_0}^t b(s)ds}{t^p}\right] = \frac{1}{t^p}\left[b(t) - \frac{p}{t}\int_{t_0}^t b(s)ds\right]$$

$$= t^{-p}\left(t^{\zeta p} - \frac{p}{1+\zeta p} \cdot \frac{t^{1+\zeta p} - t_0^{1+\zeta p}}{t}\right)$$

$$> t^{-p}\left(t^{\zeta p} - \frac{p}{1+\zeta p} \cdot t^{\zeta p}\right)$$

$$= \frac{1 - (1 - \zeta)p}{1 + \zeta p} \cdot t^{(\zeta-1)p} > 0, \quad t \geq t_0.$$

So, we get

$$d(T) = \left(\frac{\int_{t_0}^{T} b(s)ds}{T^p} \right)^{\frac{q}{p}} = \left[\frac{T^{1+\zeta p} - t_0^{1+\zeta p}}{(1 + \zeta p) \cdot T^p} \right]^{\frac{1}{p-1}}$$

$$\sim \text{constant} \cdot T^{\zeta \cdot \frac{p}{p-1} - 1} \quad \text{when } T \to +\infty$$

and respectively

$$T \cdot d(T) \sim \text{constant} \cdot T^{\zeta \cdot \frac{p}{p-1}} \quad \text{when } T \to +\infty. \tag{3.36}$$

The estimate (3.32) implies, by means of (3.36), that

$$\frac{x(t)}{t} = o\left(\left(t^{\zeta \cdot \frac{p}{p-1}} \right)^{\frac{1}{\alpha q}} \right) = o\left(t^{\frac{\zeta}{\alpha}} \right) \quad \text{as } t \to +\infty. \tag{3.37}$$

On the other hand, by taking $x_0 > 0$, a direct computation of the solution to (3.35) yields

$$\frac{[x(t)]^{(1+\varepsilon)\alpha} - x_0^{(1+\varepsilon)\alpha}}{(1 + \varepsilon)\alpha} = \frac{t^{\zeta+(1+\varepsilon)\alpha} - t_0^{\zeta+(1+\varepsilon)\alpha}}{\zeta + (1 + \varepsilon)\alpha}$$

and respectively

$$x(t) \sim \text{constant} \cdot t^{\frac{\zeta+(1+\varepsilon)\alpha}{(1+\varepsilon)\alpha}}$$

$$= \text{constant} \cdot t^{1 + \frac{\zeta}{(1+\varepsilon)\alpha}} \quad \text{when } t \to +\infty.$$

This computation shows that

$$\frac{x(t)}{t} = O\left(t^{\frac{1}{1+\varepsilon} \cdot \frac{\zeta}{\alpha}} \right) = o\left(t^{\frac{\zeta}{\alpha}} \right) \quad \text{when } t \to +\infty. \tag{3.38}$$

It is clear now that (3.38) is "infinitesimally close" to (3.37) when $\varepsilon \sim 0$.

Our intention in the remaining of this section is to produce a fractional variant of (3.32). To this end, set $\alpha \in (0, 1)$ and consider the initial value problem

$$\begin{cases} {}_0\mathbb{D}_t^\alpha(x)(t) + f(t, x(t)) = 0, \, t > 0, \\ x(0) = x_0 \in \mathbb{R}, \end{cases} \tag{3.39}$$

where the function $f : \mathbb{R}_+ \times \mathbb{R} \to \mathbb{R}$ is presumed continuous and $f(t, 0) = 0$ everywhere in $\mathbb{R}_+$. Recall that the approximate Caputo differential operator ${}_0\mathbb{D}_t^\alpha$ has been defined at page 22. The integral representation of this problem is provided by (2.2). However, given the generality of the nonlinearity f, allow us to replace the quantity $\frac{1}{\Gamma(\alpha)} \cdot f$ in there simply with f. That is, we shall work in the sequel with the integral equation

$$x(t) = x_0 - \int_0^t \frac{f(s, x(s))}{(t-s)^{1-\alpha}} ds, \quad t \geq 0. \tag{3.40}$$

Its solutions $x \in C(\mathbb{R}_+, \mathbb{R})$ are the *mild solutions* of the problem (3.39).

Take $a \in (0, 1)$ and $p_2 > 1$ such that $\alpha > 1 - a$ and

$$(1-a)\alpha = \frac{(1-a)(\alpha + \alpha^2)}{1+\alpha} < \frac{1}{p_2} < \frac{(1-a)\alpha + \alpha^2}{1+\alpha} < \alpha. \tag{3.41}$$

Introduce further $p_1, p_3 > 1$ via the formulas

$$\frac{1}{p_1} + \frac{1}{p_2} = \frac{1 + (1-a)\alpha}{1+\alpha} \quad \text{and} \quad p_3 = \frac{1+\alpha}{a\alpha}.$$

Notice that

$$\frac{1}{p_1} + \frac{1}{p_2} + \frac{1}{p_3} = 1, \tag{3.42}$$

which makes p_1, p_2, p_3 perfect candidates for the triple interpolation inequality on the real line, see [Gilbarg and Trudinger (1998), p. 146]. For further use, remark that the double inequality regarding p_2 from (3.41) is equivalent with

$$1 - \alpha < \frac{1}{p_1} < \frac{1 - (1-a)\alpha^2}{1+\alpha}. \tag{3.43}$$

Consider now the continuous, non–decreasing function $g : [0, +\infty) \to [0, +\infty)$ with the following features

$$g(0) = 0, \quad g(\xi) > 0 \text{ when } \xi > 0,$$

and — take $x_0 \neq 0$ in (3.40) —

$$\lim_{u \to +\infty} \frac{W(u)}{u^a} = +\infty, \quad W(u) = \int_0^u \frac{d\xi}{\left[g\left(|x_0| + \xi^{\frac{1}{p_3}}\right)\right]^{p_3}}. \tag{3.44}$$

An example of g is that of $g(u) = u^{1-(1+\zeta)a}$ for some $\zeta \in (0, 1)$ and $a \in \left(0, \frac{1}{1+\zeta}\right)$. It is easy to see that

$$W(u) = \int_0^u \frac{d\xi}{\left[\left(|x_0| + \xi^{\frac{1}{p_3}}\right)^{1-(1+\zeta)a}\right]^{p_3}} \geq \int_{|x_0|^{p_3}}^u \frac{d\xi}{\left[\left(2 \cdot \xi^{\frac{1}{p_3}}\right)^{1-(1+\zeta)a}\right]^{p_3}}$$

$$= 2^{-[1-(1+\zeta)a]p_3} \cdot \frac{u^{(1+\zeta)a} - |x_0|^{(1+\zeta)ap_3}}{(1+\zeta)a}$$

$$\sim c \cdot u^{(1+\zeta)a} \quad \text{when } u \to +\infty,$$

where $c = 2^{-[1-(1+\zeta)a]p_3} \cdot \frac{1}{(1+\zeta)a}$. A crucial fact should be emphasized at this point. We shall use in the following the function g given by $g(u) = u^\lambda$ for $\alpha > 1 - a > \lambda$. This restriction upon λ, which can be recast as $1 - \lambda > a$, is natural in the above computation since we can take $(1 + \zeta)a = 1 - \lambda$ for a ζ small enough.

We introduce next the class of Bihari–like[3] nonlinearities f by means of the inequality

$$|f(t,x)| \leq h(t)g\left(\frac{|x|}{(t+1)^\alpha}\right),\qquad(3.45)$$

where the function $h : [0, +\infty) \to [0, +\infty)$ is continuous and such that

$$t^{\frac{p_3}{p_1}[1-p_1(1-\alpha)]} \cdot \|h\|_{L^{p_2}((0,t),\mathbb{R})}^{p_3} = O(t^\alpha) \quad \text{when } t \to +\infty.\qquad(3.46)$$

The ratio inside g from (3.45) is designed, following the tradition of [Bihari (1957), pp. 277–278], according to our aim of studying the so called *long-time behavior* (that is, for $t \to +\infty$) of solutions to (3.39). This is why the factor $(t + 1)^\alpha$ from the argument of g looks sloppy when t is close to 0. A careful inspection of the computations will allow the interested reader to adapt the description of f for the vicinity of 0 to finer estimates.

Our main interest here is about fractional differential equations where the nonlinearity $f(t, x)$ is *power-like*. This is motivated by many applied examples: logistic equations used in population dynamics and biochemistry (e.g., polymer growth), Emden-Fowler and Thomas-Fermi type of equations (from stellar dynamics where the chaotic expansions of the outer space gases might be modeled with fractional differential equations). The architecture of the integral W from equation (3.44) has been chosen, however, as to give a certain freedom to the nonlinearity. One can have even some non-deterministic/non-smooth alterations of the power-like form.

The restriction (3.46) can be recast as

$$t^{\frac{p_3}{p_1}[1-p_1(1-\alpha)]} \cdot \left\{\int_0^t [h(s)]^{p_2}\,ds\right\}^{\frac{p_3}{p_2}} \leq M(t+1)^\alpha, \quad t \geq 0,\qquad(3.47)$$

for some sufficiently large constant M.

Notice the inequality of the exponents of t, namely

$$\frac{p_3}{p_1}[1 - p_1(1 - \alpha)] < \alpha,$$

---

[3]See the discussion at page 76.

which is equivalent with the second part of the double inequality (3.43). This means that the class of coefficients h is quite substantial, including, for instance, the space $L^{p_2}((0, +\infty), \mathbb{R})$. The computations in this section allow us to discuss the case of *slowly-decaying* coefficients h, e.g., $h(t) \sim t^{-\varepsilon}$ for some $\varepsilon \in (0, 1)$ when $t \to +\infty$.

Theorem 3.3. ([Baleanu and Mustafa (2009), Theorem 1]) *Assume that the function f from (3.40) verifies the conditions (3.45), (3.47). Then, all the mild solutions to the problem (3.39) have the asymptotic behavior*

$$x(t) = o(t^{a\alpha}) \quad as \ t \to +\infty. \tag{3.48}$$

Proof. We split the demonstration into two steps.

In *step 1*, we shall establish the raw asymptotic description of solutions to (3.39) given by the formula

$$x(t) = o(t^{\alpha}) \quad \text{when } t \to +\infty. \tag{3.49}$$

To this end, consider x a continuous solution of the integral equation (3.40) and set $y(t) = x(t) - x_0$ for every $t \geq 0$.

We have the estimates

$$|y(t)| \leq \int_0^t \frac{1}{(t-s)^{1-\alpha}} \cdot h(s) \cdot g\left(\frac{|y(s)| + |x_0|}{(s+1)^\alpha}\right) ds$$

$$\leq \left\{\int_0^t \left[\frac{1}{(t-s)^{1-\alpha}}\right]^{p_1}\right\}^{\frac{1}{p_1}} \cdot \|h\|_{L^{p_2}((0,t),\mathbb{R})} \cdot [z(t)]^{\frac{1}{p_3}}, \quad t > 0,$$

where

$$z(t) = \int_0^t \left[g\left(\frac{|y(s)| + |x_0|}{(s+1)^\alpha}\right)\right]^{p_3} ds.$$

Further — recall (3.47) —,

$$|y(t)|^{p_3} \leq c \cdot t^{\frac{p_3}{p_1}[1-p_1(1-\alpha)]} \|h\|_{L^{p_2}((0,t),\mathbb{R})}^{p_3} \cdot z(t)$$

$$\leq M_1(t+1)^\alpha \cdot z(t), \quad t \geq 0, \tag{3.50}$$

with $c = [1 - p_1(1-\alpha)]^{-\frac{p_3}{p_1}}$ and $M_1 = c \cdot M$.

We deduce now that

$$z'(t) \leq [g(|y(t)| + |x_0|)]^{p_3} \leq g\left(|x_0| + [M_1(t+1)^\alpha \cdot z(t)]^{\frac{1}{p_3}}\right)^{p_3}.$$

Fix $t_0 > 0$. Notice that, when $t \in [0, t_0]$, the preceding inequality reads as

$$z'(t) \leq g\left(|x_0| + [c_1 \cdot z(t)]^{\frac{1}{p_3}}\right)^{p_3}, \quad c_1 = M_1(1+t_0)^\alpha,$$

and

$$\frac{[c_1 z(t)]'}{g\left(|x_0| + [c_1 z(t)]^{\frac{1}{p_3}}\right)^{p_3}} \leq c_1.$$

An integration with respect to t leads to

$$W(c_1 z(t)) \leq c_1 t \leq c_1 t_0 < M_1 (1 + t_0)^{\alpha+1}, \quad t \in [0, t_0]. \qquad (3.51)$$

Replacing t with t_0 in (3.51), we get — recall (3.50) —

$$W(|y(t)|^{p_3}) \leq W(M_1(t+1)^\alpha z(t)) \leq M_1(t+1)^{\alpha+1}, \quad t \geq 0. \qquad (3.52)$$

The inequality (3.52) can be rewritten as

$$\frac{W(|y(t)|^{p_3})}{[|y(t)|^{p_3}]^a} \cdot \left[\frac{|y(t)|}{(t+1)^\alpha}\right]^{ap_3} \cdot \frac{(t+1)^{a\alpha p_3}}{(t+1)^{\alpha+1}} \leq M_1. \qquad (3.53)$$

Notice that the third factor from the second–hand side of (3.53) is 1, that is

$$\frac{W(|y(t)|^{p_3})}{[|y(t)|^{p_3}]^a} \cdot \left[\frac{|y(t)|}{(t+1)^\alpha}\right]^{ap_3} \leq M_1, \quad t \geq 0. \qquad (3.54)$$

Suppose now, for the sake of contradiction, that $y(t)$ is not $o(t^\alpha)$ when $t \to +\infty$. This means that there exist an increasing, unbounded from above, sequence of positive numbers $(t_n)_{n \geq 1}$ and $\varepsilon > 0$ such that

$$\frac{|y(t_n)|}{(t_n)^\alpha} \geq \varepsilon, \quad n \geq 1. \qquad (3.55)$$

Since (3.55) implies that $\lim_{n \to +\infty} |y(t_n)| = +\infty$, we have — recall (3.44)

$$\lim_{n \to +\infty} \frac{W(|y(t_n)|^{p_3})}{[|y(t_n)|^{p_3}]^a} = \lim_{u \to +\infty} \frac{W(u)}{u^a} = +\infty.$$

According to the inequality (3.54), we obtain that $\lim_{n \to +\infty} \frac{|y(t_n)|}{(t_n)^\alpha} = 0$ and this contradicts (3.55).

Step 2. According to the first step, $y(t) = o(t^\alpha)$ for all the large values of t. This yields, via the L'Hôpital rule, that

$$\lim_{t \to +\infty} \frac{z(t)}{t} = \lim_{t \to +\infty} \frac{\int_0^t \left[g\left(\frac{|y(s)|+|x_0|}{(s+1)^\alpha}\right)\right]^{p_3} ds}{t} = \lim_{t \to +\infty} g\left(\frac{|y(t)|+|x_0|}{(t+1)^\alpha}\right)$$
$$= g(0) = 0.$$

Further, by means of the estimate (3.50), we conclude that

$$|y(t)| = o\left(t^{\frac{1+\alpha}{p_3}}\right) = o(t^{a\alpha}) \quad \text{when } t \to +\infty.$$

The proof is complete. □

To illustrate the result, let us fix $\alpha, \lambda \in (0,1)$ with $\alpha > \lambda$. There exists $\zeta \in (0,1)$ with the property that

$$\alpha > \frac{\lambda + \zeta}{1 + \zeta}. \tag{3.56}$$

Set also $a = a(\zeta) = \frac{1-\lambda}{1+\zeta}$.

The number $p_2 > 1$ satisfying (3.41) can now be introduced. Noticing that the function $\zeta \mapsto \frac{\lambda+\zeta}{1+\zeta} = 1 - a(\zeta)$ is increasing, the conditions (3.41) can be simplified using the stronger restriction

$$\lambda\alpha < \frac{\lambda + \zeta}{1 + \zeta} \cdot \alpha < \frac{1}{p_2} < \lambda\alpha + (1-\lambda)\frac{\alpha^2}{1+\alpha} = \frac{\lambda\alpha + \alpha^2}{1+\alpha} \tag{3.57}$$

$$< \frac{\frac{\lambda+\zeta}{1+\zeta} \cdot \alpha + \alpha^2}{1+\alpha}.$$

In fact, let us fix p_2 such that

$$\lambda\alpha < \frac{1}{p_2} < \lambda\alpha + (1-\lambda)\frac{\alpha^2}{1+\alpha}. \tag{3.58}$$

Then, we can replace (3.56), (3.57) with the more practical requirements for ζ given by

$$\alpha > \frac{\lambda + \zeta}{1 + \zeta}, \quad \frac{\lambda + \zeta}{1 + \zeta} \cdot \alpha < \frac{1}{p_2}.$$

Further, the *data regarding the numbers* a, $(p_i)_{i \in \overline{1,3}}$ can be summarized in the following algorithm-like structure:

Step 1   Take α, λ such that

$$1 > \alpha > \lambda > 0;$$

Step 2   Take p_2 such that

$$\left[\lambda\alpha + (1-\lambda)\frac{\alpha^2}{1+\alpha}\right]^{-1} < p_2 < (\lambda\alpha)^{-1};$$

Step 3   Take ζ such that

$$0 < \zeta < \min\left\{\frac{\alpha - \lambda}{1 - \alpha}, \frac{1 - \lambda\alpha p_2}{\alpha p_2 - 1}\right\};$$

Step 4   Set

$$a = \frac{1 - \lambda}{1 + \zeta};$$

Step 5   Set

$$p_1 = \left[\frac{1 + (1 - a)\alpha}{1 + \alpha} - \frac{1}{p_2} \right]^{-1};$$

Step 6   Set

$$p_3 = \frac{1 + \alpha}{a\alpha}.$$

Consider the initial value problem for a fractional (general) logistic equation with sublinear exponents

$$\begin{cases} {}_0\mathbb{D}_t^\alpha(x)(t) = H(t)[x(t)]^\lambda \left\{ 1 + P(t)[x(t)]^{-\mu} \right\}, \, t > 0, \\ x(0) = x_0, \end{cases} \quad (3.59)$$

where $\mu \in (0, \lambda)$, the function $P : [0, +\infty) \to \mathbb{R}$ is continuous and bounded, and

$$H(t) = (t + 1)^{-\alpha\lambda} h(t), \quad h \in L^{p_2}((0, +\infty), \mathbb{R}).$$

Then, all the positive-valued, continuable (mild) solutions of the initial value problem have the asymptotic behavior $x(t) = o(t^\alpha)$ when $t \to +\infty$, according to the first step of Theorem 3.3. This *lack of sharpness* in the asymptotic estimate is explained in the next paragraph.

Notice that

$$\left| H(t)x^\lambda \left[1 + P(t)x^{-\mu} \right] \right| \leq (t + 1)^{-\alpha\lambda} h(t) \cdot (1 + \|P\|_{L^\infty((0,+\infty),\mathbb{R})})$$
$$\times (x^\lambda + x^\eta)$$
$$\leq (1 + \|P\|_{L^\infty((0,+\infty),\mathbb{R})}) h(t) \cdot 2 \left[\frac{1 + x}{(t + 1)^\alpha} \right]^\lambda$$
$$= h(t) g \left(\frac{x}{(t + 1)^\alpha} \right), \quad t \geq 0, \, x > 0,$$

where $\eta = \lambda - \mu \in (0, 1)$ and

$$g(u) = 2(1 + \|P\|_{L^\infty((0,+\infty),\mathbb{R})})(1 + u)^\lambda.$$

Since this function g does not verify the hypothesis $g(0) = 0$, we can extract from the inferences of Theorem 3.3 only the conclusion of the first step.

Let us observe also that for a, λ and p_2 subjected to the preceding restrictions, that is $0 < \lambda < \alpha < 1$ and p_2 confined to the conditions in (3.58), all the continuable (mild) solutions of the sublinear fractional differential equation

$$\begin{cases} {}_0\mathbb{D}_t^\alpha(x)(t) = H(t)[x(t)]^\lambda, \, t > 0, \\ x(0) = x_0, \end{cases} \quad (3.60)$$

where

$$H(t) = (t+1)^{-\alpha\lambda}h(t), \quad h \in C(\mathbb{R}_+, \mathbb{R}) \cap L^{p_2}((0, +\infty), \mathbb{R}),$$

have the asymptotic behavior $x(t) = o(t^{a\alpha})$ when $t \to +\infty$, according to Theorem 3.3. The superlinear variant of this equation, namely the case of (3.59) with $P \equiv 0$, $\alpha \in \left(\frac{1}{2}, 1\right)$ and $\lambda > 1$ has been discussed in [Furati and Tatar (2005)], however, under a *quite demanding restriction* on the coefficient h — for instance, $h \in L^p((0, +\infty), \mathbb{R})$ for all $p \in (0, 1)$. Such a restriction excludes any analysis of the long time behavior of solutions in the case of slowly-decaying coefficients. On the formal level, a solution of type $O(t^\alpha)$ as $t \to +\infty$ for $\alpha = \frac{1}{2}$ and $\lambda = 4$ has been presented in [Podlubny (1999a), p. 236] for a fractional model of radiative cooling of a semi-infinite body.

We shall deal now with the crucial issue of *sharpness* for our asymptotic estimate in Theorem 3.3. To this end, we build an "extreme" example[4] and show that, with respect to it, our theorem does the best job possible.

Before proceeding with the computations, let us mention that, similarly to the case of the classical Bihari investigation [Agarwal et al. (2007a)], we obtain here the following *type of error*: for an exact solution having the numerical values described by $O(t^{\varepsilon \cdot q})$ when $t \to +\infty$, where $\varepsilon \in (\varepsilon_0, 1)$ for a prescribed $\varepsilon_0 > 0$ and $q > 0$, the asymptotic behavior obtained using inequalities and averaging reads as $o(t^q)$ for $t \to +\infty$. Since we can set ε_0 as close to 1 as desired, it is obvious that the estimate provided by such a Bihari-type analysis is optimal.

Consider now the integral equation below

$$x(t) = x_0 + \int_0^t \frac{H(s)}{(t-s)^{1-\alpha}}[x(s)]^\lambda ds, \quad t > 0, \qquad (3.61)$$

where $x_0 > 0$ and $\alpha, \lambda \in (0, 1)$ are fixed numbers such that

$$1 > \frac{3\alpha}{2} \quad \text{and} \quad \alpha > 2\lambda. \qquad (3.62)$$

Here, $H : [0, +\infty) \to [0, +\infty)$ is a continuous function.

The next lemmas address three fundamental issues regarding the investigation of (3.61): the sign of x, its uniqueness and its global existence in the future.

Lemma 3.1. *If $x : [0, T) \to \mathbb{R}$, where $T \leq +\infty$, is a continuous solution of equation (3.61) such that $x(0) = x_0$ then*

$$x(t) \geq x_0, \quad t \in [0, T). \qquad (3.63)$$

---

[4]Recall (3.35).

Proof. Notice that it is enough to prove that x takes positive values everywhere in $[0, T)$. In fact, if we assume that x has a zero in $[0, T)$ then, given its continuity, there exists the number $t_0 > 0$ such that — recall that $x(0) = x_0 > 0$ —

$$x(t) > 0 \quad \text{for all } t \in [0, t_0), \quad x(t_0) = 0.$$

By taking into account (3.61), we get

$$x(t_0) = x_0 + \int_0^{t_0} \frac{H(s)}{(t_0 - s)^{1-\alpha}} [x(s)]^\lambda ds \geq x_0.$$

We have reached a contradiction.

Finally, since $x(t) > 0$ throughout $[0, T)$, we obtain (3.63). $\square$

Lemma 3.2. *Given the number $x_0 > 0$, there exists at most one continuous solution $x : [0, T) \to \mathbb{R}$, where $T \leq +\infty$, of equation (3.61) such that $x(0) = x_0$.*

Proof. Suppose that there exists two continuous solutions $x, y : [0, T) \to \mathbb{R}$ of equation (3.61) with $x(0) = y(0) = x_0$. Then, via Lemma 3.1, we have

$$x(t), \, y(t) \geq x_0 \quad \text{for all } t \in [0, T).$$

The mean value theorem, applied to the function $k : \left(\frac{x_0}{2}, +\infty\right) \to (0, +\infty)$ with the formula $k(\xi) = \xi^\lambda$, yields

$$|x^\lambda - y^\lambda| = \left| \frac{\lambda}{\zeta^{1-\lambda}} \cdot (\max\{x, y\} - \min\{x, y\}) \right| \leq \lambda \left(\frac{2}{x_0} \right)^{1-\lambda} \cdot |x - y|,$$

where ξ lies between x and y.

The latter estimate implies that

$$|x(t) - y(t)| \leq \int_0^t \frac{H(s)}{(t - s)^{1-\alpha}} \cdot |[x(s)]^\lambda - [y(s)]^\lambda| ds$$

$$\leq c \cdot \int_0^t \frac{H(s)}{(t - s)^{1-\alpha}} \cdot |x(s) - y(s)| ds, \, t \in [0, T), \quad (3.64)$$

where $c = \lambda (2x_0^{-1})^{1-\lambda}$.

The Lipschitz-type formula (3.64) allows one to apply the iterative procedure described in [Podlubny (1999a), pp. 127–131] to conclude that the solutions x and y coincide.

Another variant is to prove that $x = y$ on an interval of *arbitrary* length, say $[0, T^\star]$, where $T^\star > 0$. Take $k = k(T^\star) = \sup_{s \in [0, T^\star]} |H(s)|$ and $\delta \in (0, 1 - \alpha)$. As $x, y \in C([0, T^\star], \mathbb{R})$, we have, obviously, $x, y \in \mathcal{RL}^\delta$. Further, by dividing the interval $[0, T^\star]$ in a finite family of closed intervals of convenient length, we may employ the technique from (2.26), (2.27) for x replaced in there with $x - y$. $\square$

Lemma 3.3. *Given the number $x_0 > 0$, there exists at least one continuous solution $x : \mathbb{R}_+ \to \mathbb{R}$ of equation (3.61) such that $x(0) = x_0$.*

Proof. The argument is based on the technique developed in Theorem 2.1. Here, $f(t, x) = \Gamma(\alpha) \cdot H(t)x^\lambda$ and $F(t) = c \cdot H(t)$, with $c = \lambda(2x_0^{-1})^{1-\lambda}$ (it has been defined in the preceding demonstration).

Allow us to refer to the λ in the proof of Theorem 2.1 as L. The only modification we have to operate in that proof consists of renewing the definition of the (complete) metric space. Namely, we have to use $\mathcal{X} = (X, d_L)$, where X is the set of all elements of $C(\mathbb{R}_+, [x_0, +\infty))$ that behave as $O(H_L(t))$ when t goes to $+\infty$. $\square$

Consider further the particular case of equation (3.61) given by

$$H(t) = (t + 1)^{-\alpha\lambda} h(t) = (t + 1)^{-\alpha\lambda} \cdot t^{-\frac{1+\varepsilon}{p_2}}, \quad t \geq 1,$$

where $\varepsilon \in (0, 1)$ is small enough to have $(1 - \lambda)\alpha > \frac{1+\varepsilon}{p_2}$. The existence of ε will be explained in the next paragraph. As regards the definition of H, this function is continued downward to $t = 0$ in such a way as to keep its sign and continuity in $[0, +\infty)$.

It is supposed that the number p_2 verifies the double inequality (3.58). To see how to choose ε, notice that the inequality below

$$(1 - \lambda)\alpha > \lambda\alpha + (1 - \lambda)\frac{\alpha^2}{1 + \alpha} \quad \left(> \frac{1}{p_2}\right)$$

is equivalent with

$$1 > \lambda(2 + \alpha).$$

The hypotheses (3.62) lead to

$$1 > \frac{3\alpha}{2} > \alpha + \frac{\alpha^2}{2} = \frac{\alpha}{2} \cdot (2 + \alpha) > \lambda \cdot (2 + \alpha),$$

so such a number ε exists truly.

Let now x be the solution of equation (3.61) given by Lemmas 3.1, 3.2, 3.3 for $T = +\infty$ — such a solution exists always via Theorem 2.1 — . By taking into account the formula of H, we obtain that

$$x(t) \geq x_0 + \int_1^t \frac{h(s)}{(t - s)^{1-\alpha}} \cdot \left[\frac{x(s)}{(s + 1)^\alpha}\right]^\lambda ds$$

$$\geq x_0 + \int_1^t \frac{s^{-\frac{1+\varepsilon}{p_3}}}{(t-s)^{1-\alpha}} \cdot \left[\frac{x_0}{(s+1)^\alpha}\right]^\lambda ds$$

$$\geq \int_1^t \frac{ds}{(t-s)^{1-\alpha}} \cdot \frac{1}{t^{\frac{1+\varepsilon}{p_3}}} \cdot \frac{x_0^\lambda}{(t+1)^{\alpha\lambda}}$$

$$\sim c \cdot t^{(1-\lambda)\alpha - \frac{1+\varepsilon}{p_3}} \quad \text{when } t \to +\infty, \tag{3.65}$$

where $c = \frac{x_0^\lambda}{\alpha}$.

According to our theorem, $x(t) = o(t^{\alpha a})$ for $t \to +\infty$, where $\lambda < 1 - a < \alpha$. Since we can fix a in this range freely, assume that a is taken such that

$$\alpha > 2\lambda > 1 - a > \lambda. \tag{3.66}$$

This choice is in perfect agreement with the hypotheses (3.62).

The restriction (3.66) leads to — recall (3.58) —

$$\frac{1}{p_2} > \lambda\alpha > (1 - \lambda - a)\alpha.$$

Further, we have

$$a\alpha > (1 - \lambda)\alpha - \frac{1}{p_2} > (1 - \lambda)\alpha - \frac{1+\varepsilon}{p_2} > 0.$$

We would like to evaluate the difference between the exponent $a\alpha$ of t in the asymptotic formula of the solution x given by our theorem and the exponent $(1 - \lambda)\alpha - \frac{1+\varepsilon}{p_2}$ provided by (3.65). This reads as

$$a\alpha - \left[(1 - \lambda)\alpha - \frac{1+\varepsilon}{p_2}\right] = \frac{1+\varepsilon}{p_2} + (a + \lambda - 1)\alpha$$

$$< (1 + \varepsilon)\left[\lambda\alpha + (1 - \lambda)\frac{\alpha^2}{1+\alpha}\right] + (a + \lambda - 1)\alpha$$

$$= \alpha\left\{[a + (2 + \varepsilon)\lambda - 1] + \frac{1+\varepsilon}{1+\alpha}[\alpha(1 - \lambda)]\right\}$$

$$< \alpha\{[a + (2 + \varepsilon)\lambda - 1] + (1 + \varepsilon)\alpha\} = \Omega\alpha.$$

Fix now $\eta_0 \in (0, 1)$ and replace the first of hypotheses (3.62) with a stronger condition, namely

$$\eta_0 > \frac{7\alpha}{2}.$$

Thus, we have

$$\Omega < \eta_0 a.$$

In fact, this inequality is equivalent with

$$(1 - \eta_0)a + (2 + \varepsilon)\lambda + (1 + \varepsilon)\alpha < 1.$$

The latter inequality is valid as a consequence of the estimate below — recall the second of hypotheses (3.62) —

$$(1 - \eta_0)a + (2 + \varepsilon)\lambda + (1 + \varepsilon)\alpha < (1 - \eta_0) + 3\lambda + 2\alpha$$
$$< 1 - \eta_0 + \frac{3\alpha}{2} + 2\alpha < 1.$$

The conclusion of this analysis of equation (3.61) is that the solution x, if computed numerically, will be at least as big as $O(t^{(1-\eta)\cdot a\alpha})$ for $t \to +\infty$, where $\eta \in (0, \eta_0)$ and η_0 is a *prescribed* quantity, while the description of x given by our theorem reads as $o(t^{a\alpha})$ when $t \to +\infty$. In other words, the theorem provides a sharp asymptotic estimate for the mild solutions of (3.39) and, in some particular cases, this estimate is optimal.

3.4 The Bihari Asymptotic Integration Theory of the Differential Equations of Second Order

Consider the nonlinear ordinary differential equation

$$x''(t) = a(t)w\left(\frac{x(t)}{t}\right), \quad t \geq t_0 \geq 1, \tag{3.67}$$

where the continuous coefficient $a : [t_0, +\infty) \to \mathbb{R}$ verifies the restriction

$$\int_{t_0}^{+\infty} |a(t)|dt < +\infty \tag{3.68}$$

and the nonlinearity w is subjected to the same restrictions as in the preceding section.

We decided to include in this chapter a sketch of a general technique that proved extremely useful in the classical asymptotic integration theories for ordinary differential equations, see the presentation in [Agarwal et al. (2007b)]. In this way, the reader can evaluate more easily the "pitfalls and challenges" of any asymptotic integration theory designed for the order *one and something*.

Taking into account (3.29), we shall discuss the following two cases:

$$(i) \int_0^1 \frac{du}{w(u)} = +\infty \quad \text{and} \quad (ii) \int_1^{+\infty} \frac{du}{w(u)} = +\infty. \tag{3.69}$$

The quantity 1 from the integrals in (3.69) can be replaced conveniently with any other positive number — recall the discussion regarding (3.19) —.

We claim that *in the case (ii), all the positive continuable solutions* x *of (3.67) are described asymptotically by*

$$\lim_{t \to +\infty} x'(t) = \lim_{t \to +\infty} \frac{x(t)}{t} = l_x \in [0, +\infty) \tag{3.70}$$

and there exist solutions with $l_x > 0$. *In the case (i), there exist positive solutions* x *of (3.67) which verify (3.70).*

Let x be a positive solution of (3.67) defined in $[t_1, +\infty)$, where $t_1 \geq t_0$. By two integrations, we get that

$$x'(t) \leq c_1 + \int_{t_1}^{t} |a(s)| w\left(\frac{x(s)}{s}\right) ds, \quad c_1 = |x'(t_1)|,$$

and

$$x(t) \leq c_2 + c_1(t - t_1) + \int_{t_1}^{t} (t - s)|a(s)| w\left(\frac{x(s)}{s}\right) ds, \quad c_2 = x(t_1) > 0.$$

The latter estimate leads to — recall that $t_0 \geq 1$ —

$$\frac{x(t)}{t} \leq c_1 + c_2 + \int_{t_1}^{t} |a(s)| w\left(\frac{x(s)}{s}\right) ds, \quad t \geq t_1. \tag{3.71}$$

According to Bihari's direct inequality (3.9), (3.11) for $c = c_1 + c_2$, $g = w$ and $k = 1$, we arrive at

$$\frac{x(t)}{t} \leq G^{-1}\left(G(c) + \int_{t_1}^{+\infty} |a(s)| ds\right) = X = X(c, t_1) < +\infty, \quad t \geq t_1.$$

Now, notice that

$$0 \leq \int_{t_1}^{t} \left| a(s) w\left(\frac{x(s)}{s}\right)\right| ds \leq \int_{t_1}^{t} |a(s)| \cdot w(X) \, ds$$

$$\leq w(X) \cdot \int_{t_1}^{+\infty} |a(s)| ds. \tag{3.72}$$

In particular, we deduce that the integral

$$\int_{t_1}^{+\infty} a(s) w\left(\frac{x(s)}{s}\right) ds$$

is convergent.

Consequently, we have

$$\lim_{t \to +\infty} x'(t) = \lim_{t \to +\infty} \left[x'(t_1) + \int_{t_1}^{t} a(s) w\left(\frac{x(s)}{s}\right) ds\right]$$

$$= x'(t_1) + \int_{t_1}^{+\infty} a(s) w\left(\frac{x(s)}{s}\right) ds = l_x. \tag{3.73}$$

The remaining part of (3.70) follows by means of L'Hôpital's rule.

Further, fix $c_1 = x'(t_1) = c_2 = d > 0$ and take $t_1 > t_0$ large enough to have

$$d > w(X) \cdot \int_{t_1}^{+\infty} |a(s)| ds.$$

This is always possible since

$$\lim_{t_1 \to +\infty} \frac{d}{w(X(2d, t_1))} = \frac{d}{w(2d)} > 0 = \lim_{t_1 \to +\infty} \int_{t_1}^{+\infty} |a(s)| ds.$$

By taking into account (3.73), we deduce that

$$l_x \geq d - \int_{t_1}^{+\infty} |a(s)| \cdot w(X) \, ds > 0.$$

In conclusion, there exist positive solutions x of (3.67) with $l_x > 0$.

Moving to the case (i), set $k > 0$ small enough to have

$$\int_{t_1}^{+\infty} |a(s)| ds < \int_{2k}^{1} \frac{du}{g(u)} < +\infty.$$

Fix also $c_1 = x'(t_1) > 0$, $c_2 > 0$ such that $c = c_1 + c_2 \in (k, 2k)$. We have

$$G(c) + \int_{t_1}^{+\infty} |a(s)| ds < G(2k) + \int_{2k}^{1} \frac{du}{g(u)} = G(1)$$

and so the inequality (3.71) implies that

$$\frac{x(t)}{t} \leq X < 1, \quad t \geq t_1.$$

The conclusion is reached in the same way as before.

In a slightly different formulation, the analysis presented in this section is due to Bihari [Bihari (1957), pp. 277–278]. In [Mustafa and Rogovchenko (2002), pp. 360–361] several examples are given of equations (3.67) in the case (i) which posses solutions x not satisfying (3.70). The existence of positive solutions x with $\liminf_{t \to +\infty} x(t) > 0$ and $l_x = 0$ of equation (3.67) is discussed in [Mustafa (2008), p. 159].

Chapter 4

Asymptotic Integration for the Differential Equations of Order $1 + \alpha$

We shall focus here on establishing fractional variants for two major issues in the classical asymptotic integration paradigm: the Trench asymptotic development [Trench (1963)] for the solutions of a differential equation as $t \to +\infty$ and the set of hypotheses needed for the absence of non-trivial L^p-solutions, which is the heart of the limit-circle/limit-point classification [Weyl (1910)] of differential operators due to H. Weyl.

4.1 An Asymptotic Integration Theory of Trench Type

Consider the linear ordinary differential equation

$$x''(t) + f(t)x(t) = 0, \quad t \geq 0,$$

where the function $f : \mathbb{R}_+ \to \mathbb{R}$ is continuous and x_1, x_2 are two linearly independent solutions that generate its solution space[1]. If $g : \mathbb{R}_+ \to \mathbb{R}$ is continuous and such that

$$\int_0^{+\infty} u(t)|g(t)|dt < +\infty, \quad u(t) = \max\{|x_1(t)|^2, |x_2(t)|^2\},$$

then for any real numbers a, b the perturbed differential equation

$$x'' + [f(t) + g(t)]x = 0, \quad t \geq 0, \tag{4.1}$$

possesses a solution $y(t)$ with the *prescribed* asymptotic development

$$y(t) = [a + o(1)] \cdot x_1(t) + [b + o(1)] \cdot x_2(t) \quad \text{when } t \to +\infty. \tag{4.2}$$

---

[1]To quickly review his/her knowledge on classical ordinary differential equations and their oscillation, asymptotic integration and spectral analysis techniques, we recommend warmly to the interested reader the monograph [Eastham (1970)].

This sharp result has been established by Trench [Trench (1963)] in 1963. Similar conclusions have been obtained in the case of bounded (for $f(t) = 1$) and asymptotically linear (for $f(t) = 0$) solutions by Wintner and Hartman [Hartman and Wintner (1955); Wintner (1947)] and in the case of nonlinear perturbations by Waltman [Waltman (1964)]. A detailed presentation of such estimates for general ordinary differential equations can be found in the section referring to the *Kusano-Trench theory* in [Agarwal et al. (2007b)].

Let us introduce the $(1 + \alpha)$–order fractional differential equation

$$\,_0^2 \mathcal{O}_t^{1+\alpha}(x)(t) + a(t)x(t) = 0, \quad t > 0, \tag{4.3}$$

where the function $a : [0, +\infty) \to \mathbb{R}$ is presumed continuous[2]. As many times before, we shall be concerned only with the *mild solutions* of (4.3), that is with those $x \in C((0, +\infty), \mathbb{R})$ verifying its integral counterpart — recall (1.62) —.

The first result introduces our leading assumptions as regards a in (4.3).

Proposition 4.1. *Set $x_0 \in \mathbb{R}$. Suppose that one has*

$$\int_0^1 \frac{|a(\tau)|}{\tau^{1-\alpha}} d\tau + \int_1^{+\infty} \tau^\alpha |a(\tau)| d\tau < +\infty.$$

Then, the problem

$$\begin{cases} \,_0^2 \mathcal{O}_t^{1+\alpha}(x)(t) + a(t)x(t) = 0, \, t > 0, \\ \lim_{t \searrow 0}[t^{1-\alpha}x(t)] = x_0, \end{cases} \tag{4.4}$$

has a solution $x \in C((0, +\infty), \mathbb{R})$ such that $\lim_{t \to +\infty} (\,_0 D_t^\alpha x)(t) \in \mathbb{R}$.

Proof. Fix $T > 0$ small enough and $k \geq 0$ large enough to have (for now, $k > 0$)

$$\psi = \max\{2, T\} \cdot \int_0^T \frac{|a(\tau)|}{\tau^{1-\alpha}} d\tau + \frac{1}{k} < \Gamma(1 + \alpha).$$

Consider also the set X of all the functions $x \in C((0, +\infty), \mathbb{R})$ such that the limit $\lim_{t \searrow 0} [t^{1-\alpha}x(t)]$ is finite and $\sup_{t \geq T} \frac{|x(t)|}{t^\alpha} < +\infty$. The set, endowed with the metric d_k expressed as

$$d_k(x_1, x_2) = \max \left\{ \sup_{t \in (0,T]} \left[t^{1-\alpha} |(x_1 - x_2)(t)| \right], \right.$$

---

[2]This choice for a has been preferred not to complicate matters. In fact, the proofs work equally well when $a \in C((0, +\infty), \mathbb{R}) \cap L^\infty((0, 1), \mathbb{R})$.

$$\sup_{t \geq T} \left[\frac{|(x_1 - x_2)(t)|}{t^\alpha} \cdot \exp\left(-k \cdot A(t)\right) \right] \right\},$$

where $A(t) = \int_T^t \tau^\alpha |a(\tau)| d\tau$ and $x_1 \, x_2 \in X$, becomes a complete metric space.

Introduce further the integral operator $\mathcal{T} : X \to C((0, +\infty), \mathbb{R})$

$$\mathcal{T}(x)(t) = x_0 \cdot t^{\alpha-1} + \frac{x_1}{\Gamma(1+\alpha)} \cdot t^\alpha$$

$$- \frac{1}{\Gamma(\alpha)} \int_0^t \frac{1}{(t-s)^{1-\alpha}} \int_0^s (ax)(\tau) d\tau ds,$$

where $x \in X$ and $t > 0$. Notice that the operator is well-defined $(\mathcal{T}(X) \subseteq X)$ as a consequence of the next estimate

$$\int_0^{+\infty} |ax|(\tau) d\tau \leq \int_0^T \frac{|a(\tau)|}{\tau^{1-\alpha}} \cdot \tau^{1-\alpha} |x(\tau)| d\tau$$

$$+ \int_T^{+\infty} \tau^\alpha |a(\tau)| \cdot \exp(kA(+\infty)) \cdot \frac{|x(\tau)|}{\tau^\alpha} \exp(-kA(\tau)) d\tau$$

$$\leq \exp(kA(+\infty)) \cdot \left(\int_0^T \frac{|a(\tau)|}{\tau^{1-\alpha}} d\tau + \int_T^{+\infty} \tau^\alpha |a(\tau)| d\tau \right)$$

$$\times d_k(x, 0).$$

In fact, for any $x \in X$, we have $\lim_{t \searrow 0} \left[t^{1-\alpha} \mathcal{T}(x)(t) \right] = x_0$ and

$\lim_{t \to +\infty} {}_0 D_t^\alpha (\mathcal{T}x)(t) = x_1 - \int_0^{+\infty} (ax)(s) ds.$

For $x_1, x_2 \in X$ and $t \in (0, T]$, we get that (here, $k > 0$)

$$t^{1-\alpha} |\mathcal{T}x_1 - \mathcal{T}x_2|(t) \leq \frac{t^{1-\alpha}}{\Gamma(\alpha)} \int_0^t \frac{ds}{(t-s)^{1-\alpha}} \cdot \int_0^T \frac{|a(\tau)|}{\tau^{1-\alpha}} d\tau \cdot d_k(x_1, x_2)$$

$$\leq \frac{T}{\Gamma(1+\alpha)} \int_0^T \frac{|a(\tau)|}{\tau^{1-\alpha}} d\tau \cdot d_k(x_1, x_2). \tag{4.5}$$

Further, when $t \geq T$, we obtain that

$$\frac{|\mathcal{T}x_1 - \mathcal{T}x_2|(t)}{t^\alpha}$$

$$\leq \frac{t^{-\alpha}}{\Gamma(\alpha)} \left(\int_0^T + \int_T^t \right) \frac{1}{(t-s)^{1-\alpha}} \int_0^s |ax|(\tau) d\tau ds$$

$$\leq \frac{t^{-\alpha}}{\Gamma(\alpha)} \left[\int_0^T \frac{1}{(T-s)^{1-\alpha}} \int_0^s |ax|(\tau) d\tau ds \right.$$

$$+ \int_T^t \frac{1}{(t-s)^{1-\alpha}} \left(\int_0^T + \int_T^s \right) |ax|(\tau) d\tau ds \Bigg]$$

$$\leq \frac{t^{-\alpha}}{\Gamma(\alpha)} \left[\frac{T^\alpha + (t-T)^\alpha}{\alpha} \cdot \int_0^T \frac{|a(\tau)|}{\tau^{1-\alpha}} \right] \cdot d_k(x_1, x_2)$$

$$+ \frac{t^{-\alpha}}{\Gamma(\alpha)} \int_T^t \frac{1}{(t-s)^{1-\alpha}} \int_T^s \frac{d}{d\tau} \left[\frac{\exp(kA(\tau))}{k} \right] d\tau ds \cdot d_k(x_1, x_2)$$

$$\leq \frac{1}{\alpha \cdot \Gamma(\alpha)} \left[\left(\frac{T}{t} \right)^\alpha + \left(1 - \frac{T}{t} \right)^\alpha \right] \cdot \int_0^T \frac{|a(\tau)|}{\tau^{1-\alpha}} d\tau$$

$$+ \frac{t^{-\alpha}}{\Gamma(\alpha)} \int_T^t \frac{ds}{(t-s)^{1-\alpha}} \cdot \int_T^t \frac{d}{d\tau} \left[\frac{\exp(kA(\tau))}{k} \right] d\tau \cdot d_k(x_1, x_2)$$

$$\leq \frac{1}{\Gamma(1+\alpha)} \left[2 \int_0^T \frac{|a(\tau)|}{\tau^{1-\alpha}} d\tau + \frac{\exp(kA(t))}{k} \right] \cdot d_k(x_1, x_2). \qquad (4.6)$$

From (4.5), (4.6), we get that $d_k(\mathcal{T}x_1, \mathcal{T}x_2) \leq \frac{\psi}{\Gamma(1+\alpha)} \cdot d_k(x_1, x_2)$. The conclusion follows via the (Banach) contraction principle. $\square$

The second result is concerned with the eventually large mild solutions of the problem (4.4).

Proposition 4.2. *Set $x_1 \in \mathbb{R}$. Assume that the hypothesis of Proposition 4.1 holds true. Suppose also that there exists $T > 0$ such that*

$$\xi = \max\{1, T\} \cdot \left[\int_0^T \frac{|a(\tau)|}{\tau^{1-\alpha}} d\tau + \int_T^{+\infty} \tau^\alpha |a(\tau)| d\tau \right] < \Gamma(1+\alpha). \quad (4.7)$$

Then, the problem (4.4) has a solution $x \in C((0, +\infty), \mathbb{R})$ such that $\lim\limits_{t \to +\infty} (_0D_t^\alpha x)(t) = x_1$.

Proof. Let us consider the complete metric space (X, d_0) from the proof of Proposition 4.1. The integral operator $\mathcal{T} : X \to C((0, +\infty), \mathbb{R})$

$$\mathcal{T}(x)(t) = x_0 \cdot t^{\alpha-1} + \frac{x_1}{\Gamma(1+\alpha)} \cdot t^\alpha$$

$$+ \frac{1}{\Gamma(\alpha)} \int_0^t \frac{1}{(t-s)^{1-\alpha}} \int_s^{+\infty} (ax)(\tau) d\tau ds,$$

where $x \in X$ and $t > 0$, is well-defined and its fixed point $x_\infty \in X$ will satisfy all the claims of the statement above (provided it exists).

For $x_1, x_2 \in X$ and $t \in (0, T]$, we have the estimates

$$t^{1-\alpha} |\mathcal{T}x_1 - \mathcal{T}x_2|(t)$$

$$\leq \frac{T^{1-\alpha}}{\Gamma(\alpha)} \int_0^T \frac{ds}{(t-s)^{1-\alpha}} \cdot \left(\int_0^T + \int_T^{+\infty} \right) |ax|(\tau) d\tau$$

$$\leq \frac{T}{\Gamma(1+\alpha)} \left(\int_0^T \frac{|a(\tau)|}{\tau^{1-\alpha}} d\tau + \int_T^{+\infty} \tau^{\alpha} |a(\tau)| d\tau \right) \cdot d_0(x_1, x_2)$$

and — here, $t \geq T$ —

$$\frac{|\mathcal{T} x_1 - \mathcal{T} x_2|(t)}{t^{\alpha}}$$

$$\leq \frac{1}{\Gamma(1+\alpha)} \left(\int_0^T \frac{|a(\tau)|}{\tau^{1-\alpha}} d\tau + \int_T^{+\infty} \tau^{\alpha} |a(\tau)| d\tau \right) \cdot d_0(x_1, x_2).$$

Finally, the operator $\mathcal{T} : X \to X$ is a contraction of coefficient $k = \frac{\xi}{\Gamma(1+\alpha)}$. The proof is complete. $\square$

Our variant of a Trench-type asymptotic estimate for the solutions of the fractional differential equation (4.3) is detailed next.

Theorem 4.1. ([Baleanu et al. (2010b), Theorem 1]) *Given the real numbers* x_0, x_1 *and the continuous function* $a : [0, +\infty) \to \mathbb{R}$ *such that (4.7) holds true, assume that*

$$\int_T^{+\infty} s^{1+\alpha} |a(s)| ds < +\infty. \tag{4.8}$$

Then, the mild solution of problem (4.4) obtained at Proposition 4.2 has the asymptotic development

$$x(t) = [X_0 + O(1)] \cdot t^{\alpha-1} + [X_1 + o(1)] \cdot t^{\alpha} \tag{4.9}$$
$$= X_1 t^{\alpha} + o(t^{\alpha}) \quad \text{when } t \to +\infty,$$

with $X_0 = x_0$ *and* $X_1 = \frac{x_1}{\Gamma(1+\alpha)}$.

Proof. Let $x \in X$ be the fixed point of operator $\mathcal{T}$ from Proposition 4.2. We have the estimates

$$x(t) - x_0 \cdot t^{\alpha-1} - \frac{x_1}{\Gamma(1+\alpha)} \cdot t^{\alpha}$$

$$= \frac{1}{\Gamma(\alpha)} \int_0^t \frac{1}{(t-s)^{1-\alpha}} \left(\int_s^t + \int_t^{+\infty} \right) (ax)(\tau) d\tau ds$$

$$= \frac{1}{\Gamma(\alpha)} \int_0^t \int_0^s \frac{d\tau}{(t-\tau)^{1-\alpha}} \cdot (ax)(s) ds + \frac{t^{\alpha}}{\Gamma(1+\alpha)} \int_t^{+\infty} (ax)(\tau) d\tau$$

$$= \frac{1}{\Gamma(1+\alpha)}$$

$$\times \left\{ \int_0^t [t^\alpha - (t-s)^\alpha](ax)(s)ds + t^\alpha \int_t^{+\infty} (ax)(\tau)d\tau \right\}, \qquad (4.10)$$

where $t > 0$.

By taking $t \geq T$, notice that

$$\left| \int_t^{+\infty} (ax)(\tau)d\tau \right| \leq \frac{1}{t^\alpha} \cdot t^\alpha \int_t^{+\infty} \tau^\alpha |a(\tau)| \cdot \frac{|x(\tau)|}{\tau^\alpha} d\tau$$

$$\leq \frac{1}{t^\alpha} \cdot \int_t^{+\infty} \tau^{2\alpha} |a(\tau)| d\tau \cdot d_0(x, 0) \qquad (4.11)$$

$$= o\left(t^{-\alpha}\right) \quad \text{when } t \to +\infty. \qquad (4.12)$$

Before going further, notice that in (4.11) we have employed a weaker version of (4.8), namely

$$\int_T^{+\infty} \tau^{2\alpha} |a(\tau)| d\tau < +\infty. \qquad (4.13)$$

To take full advantage of our strong hypothesis (4.8), we make the following estimate

$$\left| t^\alpha \int_t^{+\infty} (ax)(\tau)d\tau \right| \leq t^{\alpha-1} \int_t^{+\infty} \tau^{1+\alpha} |a(\tau)| \cdot \frac{|x(\tau)|}{\tau^\alpha} d\tau$$

$$\leq t^{\alpha-1} \int_t^{+\infty} \tau^{1+\alpha} |a(\tau)| d\tau \cdot d_0(x, 0)$$

$$= o\left(t^{\alpha-1}\right) \quad \text{when } t \to +\infty. \qquad (4.14)$$

Let us make a claim. *If the continuous function $h : [T, +\infty) \to \mathbb{R}$ is in $L^1((T, +\infty), \mathbb{R})$ then $\lim_{t \to +\infty} \frac{1}{t^\varepsilon} \int_T^t s^\varepsilon h(s)ds = 0$, where $\varepsilon \in (0, 1)$.*

To prove this claim, we use integration by parts

$$\frac{1}{t^\varepsilon} \int_T^t s^\varepsilon h(s)ds = \frac{1}{t^\varepsilon} \int_T^t s^\varepsilon \left(-\int_s^{+\infty} h(\tau)d\tau \right)' ds$$

$$= \left(\frac{T}{t} \right)^\varepsilon \int_T^{+\infty} h(\tau)d\tau - \int_t^{+\infty} h(\tau)d\tau + \varepsilon \cdot \frac{\int_T^t \frac{1}{\tau^{1-\varepsilon}} \int_\tau^{+\infty} h(\xi)d\xi d\tau}{t^\varepsilon}.$$

The first two terms of the sum tend to zero as $t \to +\infty$. As for the third term, it tends to zero also by means of the L'Hôpital Rule. Our claim is established.

Returning to (4.10), we claim that *the first integral is $O(t^{\alpha-1})$ when $t \to +\infty$.*

To prove this claim, start by noticing the inequalities[3]

$$1 - \frac{s}{t} \leq \left(1 - \frac{s}{t}\right)^\alpha \leq 1 - \alpha \cdot \frac{s}{t}, \quad t^\alpha - (t - s)^\alpha \leq s^\alpha,$$

where $t \geq T$ and $s \in [0, t]$.

Now,

$$\left| \int_0^t [t^\alpha - (t - s)^\alpha](ax)(s)ds \right|$$

$$\leq t^\alpha \int_0^t \left[1 - \left(1 - \frac{s}{t}\right)^\alpha\right] |ax|(s)ds \leq t^\alpha \int_0^t \frac{s}{t} |ax|(s)ds$$

$$\leq \frac{1}{t^{1-\alpha}} \left[\int_0^T s \cdot \frac{|a(s)|}{s^{1-\alpha}} ds + \int_T^t s^{1-\alpha} \cdot s^{2\alpha} |a(s)| ds\right] \cdot d_0(x, 0) \quad (4.15)$$

$$\leq t^{\alpha-1} \cdot \left[T \int_0^T \frac{|a(s)|}{s^{1-\alpha}} ds + \int_T^{+\infty} s^{1+\alpha} |a(s)| ds\right] \cdot d_0(x, 0)$$

$$= O\left(t^{\alpha-1}\right) \quad \text{when } t \to +\infty. \tag{4.16}$$

The second claim is established.

As before, if we employ only (4.13) then, via the first claim for $\varepsilon = 1 - \alpha$ and $h(s) = s^{2\alpha} |a(s)|$, we estimate (4.15) to be smaller than

$$t^{\alpha-1} \cdot \left[T \int_0^T \frac{|a(s)|}{s^{1-\alpha}} ds\right] \cdot d_0(x, 0) + o(1) \quad \text{when } t \to +\infty. \tag{4.17}$$

Using (4.12), (4.16), the solution $x(t)$ can be expressed as

$$x(t) = t^{\alpha-1} \left\{x_0 + \frac{t^{1-\alpha}}{\Gamma(1+\alpha)} \int_0^t [t^\alpha - (t - s)^\alpha](ax)(s)ds\right\}$$

$$+ \frac{t^\alpha}{\Gamma(1+\alpha)} \left[x_1 + \int_t^{+\infty} (ax)(s)ds\right]$$

$$= t^{\alpha-1}[x_0 + O(1)] + t^\alpha \left[\frac{x_1}{\Gamma(1+\alpha)} + o(1)\right]$$

for all large values of t. We have obtained (4.9).

However, if we take into account (4.14) then things get even better, that is, the solution can be recast as

$$x(t) = t^{\alpha-1} \left\{x_0 + \frac{t^{1-\alpha}}{\Gamma(1+\alpha)} \int_0^t [t^\alpha - (t - s)^\alpha](ax)(s)ds\right.$$

---

[3]The second part of the double inequality follows at once if we observe that, given the C^1–functions $g, h : [0, 1) \to \mathbb{R}_+$ with the formulas $g(x) = (1 - x)^\alpha$ and $h(x) = 1 - \alpha x$, we have $g(0) = h(0)$ and $g'(x) \leq h'(x)$ everywhere. Thus, since $g(x) \leq h(x)$ in $[0, 1)$, the former inequality is a consequence of putting $x = \frac{s}{t}$ in the latter one. As for the other inequality, recall (1.8).

$$+ \frac{t}{\Gamma(1+\alpha)} \int_t^{+\infty} (ax)(s)ds \Big\} + t^\alpha \cdot \frac{x_1}{\Gamma(1+\alpha)}$$

$$= t^{\alpha-1}[x_0 + O(1) + o(1)] + t^\alpha \frac{x_1}{\Gamma(1+\alpha)}$$

$$= [X_0 + O(1)] \cdot t^{\alpha-1} + X_1 \cdot t^\alpha \quad \text{when } t \to +\infty. \qquad (4.18)$$

The proof is complete. $\square$

Consider further the $(1+\alpha)$–order fractional differential equation

$$_0\mathcal{O}_t^{1+\alpha}(x)(t) + a(t)x(t) = 0, \quad t > 0, \qquad (4.19)$$

where the function $a : [0, +\infty) \to \mathbb{R}$ is, again, presumed continuous. Recall as well the integral representation (1.61).

Theorem 4.2. ([Baleanu et al. (2011b), Theorem 1]) *Assume that there exists* $T > 0$ *such that* $\int_T^{+\infty} s^{1+\alpha}|a(s)|ds < +\infty$ *and*

$$k = \frac{\max\{1, T^\alpha\}}{\Gamma(1+\alpha)} \cdot \left[\int_0^T |a(s)|ds + \int_T^{+\infty} s^\alpha|a(s)|ds \right] < 1.$$

Then, given the real numbers b, c, *the differential equation (4.19) has a mild solution* $x \in C([0, +\infty), \mathbb{R})$ *with the asymptotic formula*

$$x(t) = b + c \cdot t^\alpha + O(t^{\alpha-1}) = b + ct^\alpha + o(1) \quad \text{when } t \to +\infty. \quad (4.20)$$

Proof. Let X be the set of all the functions $x \in C([0, +\infty), \mathbb{R})$ with $\sup_{t \geq T} \frac{|x(t)|}{t^\alpha} < +\infty$ and d the following metric

$$d(x_1, x_2) = \max \Big\{ \|x_1 - x_2\|_{L^\infty((0,T),\mathbb{R})},$$
$$\sup_{t \geq T} \frac{|x_1(t) - x_2(t)|}{t^\alpha} \Big\}, \quad \text{where } x_1, x_2 \in X.$$

Obviously, $\mathcal{M} = (X, d)$ is a complete metric space.

Notice that

$$\int_0^{+\infty} s^j|ax|(s)ds \leq \left[\int_0^T s^j|a(s)|ds + \int_T^{+\infty} s^{j+\alpha}|a(s)|ds \right] d(x, 0)$$
$$= C(j) \cdot d(x, 0),$$

where $j \in \{0, 1\}$, for every $x \in \mathcal{M}$.

Introduce the operator $\mathcal{T} : \mathcal{M} \to C([0, +\infty), \mathbb{R})$ with the formula

$$\mathcal{T}(x)(t) = b + ct^\alpha + \frac{1}{\Gamma(\alpha)} \int_0^t \frac{1}{(t-s)^{1-\alpha}} \int_s^{+\infty} (ax)(\tau)d\tau ds, \quad t > 0.$$

We have the estimates

$$|\mathcal{T}(x)(t)| \le |b| + |c|t^\alpha + \frac{1}{\Gamma(\alpha)} \int_0^t \frac{ds}{(t-s)^{1-\alpha}} \cdot \int_0^{+\infty} |ax|(\tau)d\tau$$

$$\le |b| + T^\alpha \left[|c| + \frac{C(0)}{\Gamma(1+\alpha)} \cdot d(x,0) \right], \quad t \in [0, T],$$

and

$$|\mathcal{T}(x)(t)| \le t^\alpha \left[\frac{|b| + T^\alpha |c|}{T^\alpha} + \frac{C(0)}{\Gamma(1+\alpha)} \cdot d(x,0) \right], \quad t \ge T,$$

which imply that $\mathcal{T}(x) \in \mathcal{M}$ and

$$d(\mathcal{T}(x), 0)$$
$$\le \max\left\{1, \frac{1}{T^\alpha}\right\}(|b| + T^\alpha |c|) + \max\{1, T^\alpha\}\frac{C(0)}{\Gamma(1+\alpha)} \cdot d(x,0),$$

where $x \in \mathcal{M}$.

We also have

$$d(\mathcal{T}(x_1), \mathcal{T}(x_2)) \le \frac{\max\{1, T^\alpha\}}{\Gamma(1+\alpha)} C(0) \cdot d(x_1, x_2), \quad x_1, x_2 \in \mathcal{M},$$

which means that $\mathcal{T} : \mathcal{M} \to \mathcal{M}$ is a contraction of coefficient k.

Let $x_0 \in \mathcal{M}$ be its fixed point. Following closely the computations from Theorem 4.1, we have the estimates

$$\int_0^t \frac{1}{(t-s)^{1-\alpha}} \int_s^{+\infty} |ax_0|(\tau)d\tau ds$$
$$= \int_0^t |ax_0|(\tau) \frac{t^\alpha - (t-\tau)^\alpha}{\alpha} d\tau + \frac{t^\alpha}{\alpha} \int_t^{+\infty} |ax_0| ds$$
$$\le \frac{t^\alpha}{\alpha} \left[\int_0^t |ax_0|(\tau) \cdot \frac{\tau}{t} d\tau + \frac{1}{t} \int_t^{+\infty} s|ax_0|(s)ds \right]$$
$$\le \frac{2C(1)}{\alpha} d(x_0, 0) \cdot t^{\alpha-1} = O(t^{\alpha-1}) \quad \text{when } t \to +\infty.$$

Finally,

$$x_0(t) = \mathcal{T}(x_0)(t) = b + ct^\alpha + O(t^{\alpha-1}) \quad \text{when } t \to +\infty.$$

The proof is complete: we have obtained (4.20). $\square$

Moving on, consider the $(1+\alpha)$–order fractional differential equation

$$\,_0^3\mathcal{O}_t^{1+\alpha}(x)(t) + a(t)x(t) = 0, \quad t > 0, \tag{4.21}$$

where the function $a : [0, +\infty) \to \mathbb{R}$ is taken continuous. Remember the integral representation (1.64).

Theorem 4.3. ([Baleanu et al. (2011b), Theorem 2]) *Suppose that*

$$\int_0^{+\infty} t|a(t)|dt + \sup_{t>0} \left[t^{1-\alpha} \int_0^t \frac{s|a(s)|}{(t-s)^{1-\alpha}} ds \right] < +\infty$$

and

$$k = \frac{1}{\Gamma(\alpha)} \left(\int_0^{+\infty} \frac{|a(s)|}{s^{1-\alpha}} ds + \chi \right) < 1,$$

where $\chi = \sup\limits_{t>0} t^{1-\alpha} \int_0^t \frac{|a(s)|}{(t-s)^{1-\alpha} s^{1-\alpha}} ds$. *Then, given the real numbers* b, c, *the differential equation (4.21) has a mild solution* $x \in C^1((0, +\infty), \mathbb{R})$ *with the asymptotic formula*

$$x(t) = [b + O(1)] \cdot t^{\alpha-1} + c \cdot t = ct + O(t^{\alpha-1}) \quad \text{when } t \to +\infty. \quad (4.22)$$

Proof. Let us start by giving a *simple example* of χ. If the functional coefficient $a \in (C \cap L^1)([0, +\infty), \mathbb{R})$ verifies the restriction

$$|a(t)| \le \frac{A}{t^\alpha}, \quad t > 0, \quad (4.23)$$

then

$$t^{1-\alpha} \int_0^{2t} \frac{|a(s)|}{(2t-s)^{1-\alpha} s^{1-\alpha}} ds = t^{1-\alpha} \left(\int_0^t + \int_t^{2t} \right) \frac{|a(s)|}{(2t-s)^{1-\alpha} s^{1-\alpha}} ds$$

$$\le t^{1-\alpha} \int_0^t \frac{|a(s)|}{t^{1-\alpha} s^{1-\alpha}} ds + t^{1-\alpha} \int_t^{2t} \frac{A}{(2t-s)^{1-\alpha} s} ds$$

$$= \left(\int_0^1 \frac{|a(s)|}{s^{1-\alpha}} ds + \int_1^{1+t} \frac{|a(s)|}{s^{1-\alpha}} ds \right) + \frac{A}{2^{1-\alpha}} \int_{\frac{1}{2}}^1 \frac{dv}{(1-v)^{1-\alpha} v}$$

$$\le \left(\int_0^1 \frac{ds}{s^{1-\alpha}} \cdot \|a\|_{L^\infty((0,1),\mathbb{R})} + \int_1^{+\infty} |a(s)| ds \right) + \frac{A}{2^{1-\alpha}} \int_{\frac{1}{2}}^1 \frac{dv}{(1-v)^{1-\alpha} \cdot \frac{1}{2}}$$

$$= \frac{1}{\alpha} \|a\|_{L^\infty((0,1),\mathbb{R})} + \|a\|_{L^1((1,+\infty),\mathbb{R})} + 2^\alpha A \cdot \frac{\left(\frac{1}{2}\right)^\alpha}{2}$$

$$= \alpha^{-1} \|a\|_{L^\infty((0,1),\mathbb{R})} + \|a\|_{L^1((1,+\infty),\mathbb{R})} + \frac{A}{2} < +\infty, \quad t > 0.$$

Notice also that $\int_0^t \frac{|a(s)|}{(t-s)^{1-\alpha}} ds \le t^{1-\alpha} \int_0^t \frac{|a(s)|}{(t-s)^{1-\alpha} s^{1-\alpha}} ds \le \chi$ and

$$t^{1-\alpha} \int_0^t \frac{s|a(s)|}{(t-s)^{1-\alpha}} ds = t^{1-\alpha} \int_0^t \frac{s^{2-\alpha}|a(s)|}{(t-s)^{1-\alpha} s^{1-\alpha}} ds, \quad t > 0,$$

which leads to the "χ" of the mapping $t \mapsto t^{2-\alpha} a(t)$ in $[0, +\infty)$.

Introduce now the set Z of all the functions $y \in C((0, +\infty), \mathbb{R})$ such that $\sup\limits_{t>0} t^{1-\alpha}|y(t)| < +\infty$ and the metric

$$d(y_1, y_2) = \sup_{t>0} t^{1-\alpha}|y_1(t) - y_2(t)|, \quad y_1, y_2 \in Z.$$

Observe also that

$$\sup_{t>0} t^{2-\alpha} \int_t^{+\infty} \frac{|y_1(u) - y_2(u)|}{u^2} du \leq \frac{1}{2-\alpha} \cdot \sup_{t>0} t^{1-\alpha}|y_1(t) - y_2(t)|$$

$$\leq d(y_1, y_2). \tag{4.24}$$

The metric space $\mathcal{P} = (Z, d)$ is complete.

Further, define the integral operator $\mathcal{T} : \mathcal{P} \to C((0, +\infty), \mathbb{R})$ via the formula

$$\mathcal{T}(y)(t)$$

$$= \frac{t^{\alpha-1}}{\Gamma(\alpha)} \left[x_1 + \int_0^{+\infty} (ax)(\tau)d\tau \right] - \frac{1}{\Gamma(\alpha)} \int_0^t \frac{(ax)(s)}{(t-s)^{1-\alpha}} ds$$

$$= t^{\alpha-1} \left[b + \frac{c}{\Gamma(\alpha)} \int_0^{+\infty} sa(s)ds \right] - \frac{c}{\Gamma(\alpha)} \int_0^t \frac{sa(s)}{(t-s)^{1-\alpha}} ds$$

$$- \frac{t^{\alpha-1}}{\Gamma(\alpha)} \int_0^{+\infty} \tau a(\tau) \int_\tau^{+\infty} \frac{y(u)}{u^2} du\, d\tau$$

$$+ \frac{1}{\Gamma(\alpha)} \int_0^t \frac{\tau a(\tau)}{(t-\tau)^{1-\alpha}} \int_\tau^{+\infty} \frac{y(u)}{u^2} du\, d\tau, \quad t > 0,$$

where $b = \frac{x_1}{\Gamma(\alpha)}$ and $x(t) = t \left[c - \int_t^{+\infty} \frac{y(\tau)}{\tau^2} d\tau \right]$.

Given $y \in \mathcal{P}$, we have the estimates

$$t^{1-\alpha}|\mathcal{T}(y)(t)|$$

$$\leq |b| + \frac{|c|}{\Gamma(\alpha)} \int_0^{+\infty} s|a(s)|ds + \frac{|c|}{\Gamma(\alpha)} \cdot \sup_{t>0} t^{1-\alpha} \int_0^t \frac{s|a(s)|}{(t-s)^{1-\alpha}} ds$$

$$+ \frac{1}{\Gamma(\alpha)} \int_0^{+\infty} \frac{|a(s)|}{s^{1-\alpha}} ds \cdot \sup_{s>0} s^{2-\alpha} \int_s^{+\infty} \frac{|y(u)|}{u^2} du \tag{4.25}$$

$$+ \frac{1}{\Gamma(\alpha)} \cdot \sup_{t>0} t^{1-\alpha} \int_0^t \frac{|a(\tau)|}{(t-\tau)^{1-\alpha}\tau^{1-\alpha}} d\tau$$

$$\times \sup_{\tau>0} \tau^{2-\alpha} \int_\tau^{+\infty} \frac{|y(u)|}{u^2} du, \quad t > 0, \tag{4.26}$$

which imply that $\mathcal{T}(\mathcal{P}) \subseteq \mathcal{P}$.

Further, taking into account (4.25), (4.26) and (4.24), we have

$$t^{1-\alpha}|\mathcal{T}(y_1)(t) - \mathcal{T}(y_2)(t)|$$

$$\leq \left[\frac{1}{\Gamma(\alpha)} \int_0^{+\infty} \frac{|a(s)|}{s^{1-\alpha}} ds + \frac{1}{\Gamma(\alpha)} \sup_{t>0} t^{1-\alpha} \int_0^t \frac{|a(\tau)|}{(t-\tau)^{1-\alpha}\tau^{1-\alpha}} d\tau \right]$$
$$\times d(y_1, y_2)$$
$$= \frac{1}{\Gamma(\alpha)} \left(\int_0^{+\infty} \frac{|a(s)|}{s^{1-\alpha}} ds + \chi \right) d(y_1, y_2), \quad t > 0,$$

where y_1, $y_2 \in \mathcal{P}$.

The operator $T : \mathcal{P} \to \mathcal{P}$ being a contraction of coefficient k, it has a fixed point y_0. Thus, since $y_0(t) = O(t^{\alpha-1})$ for large values of t, we conclude the validity of the asymptotic expansion (4.22) for the solution x given by (1.63). Notice also that

$$\lim_{t \searrow 0} t^{1-\alpha} y_0(t) = \lim_{t \searrow 0} t^{1-\alpha} T(y_0)(t)$$
$$= b + \frac{1}{\Gamma(\alpha)} \int_0^{+\infty} sa(s) \left[c - \int_s^{+\infty} \frac{y_0(u)}{u^2} du \right] ds,$$

which means that $y_0 \in \mathcal{RL}^{1-\alpha}$.

The proof is complete: we have obtained (4.22). □

4.2   Asymptotically Linear Solutions

Consider the ordinary differential equation

$$x''(t) + f(t, x(t)) = 0, \quad t \geq 1, \tag{4.27}$$

where the function $f : [1, +\infty) \times \mathbb{R} \to \mathbb{R}$ is continuous and subjected to a Bihari-type restriction[4]

$$|f(t, x)| \leq h(t) \cdot g\left(\frac{|x|}{t}\right), \quad t \geq 1, x \in \mathbb{R}.$$

Here, the functions $h : [1, +\infty) \to [0, +\infty)$ and $g : [0, +\infty) \to [0, +\infty)$ are continuous and there exists $\varepsilon \in [0, 1]$ with

$$\int_1^{+\infty} t^\varepsilon h(t) dt < +\infty.$$

Then, given $c, d \in \mathbb{R}$, the equation (4.27) has a solution x, defined in a neighborhood of $+\infty$, which is expressible as $ct + o(t)$ for $\varepsilon = 0$, as $ct + o(t^{1-\varepsilon})$ for $\varepsilon \in (0, 1)$ and, finally, as $ct + d + o(1)$ for $\varepsilon = 1$ when $t \to +\infty$. Such a solution is called *asymptotically linear* in the literature. In

---

[4]Recall the discussion starting at page 76.

particular, these developments apply to the homogeneous linear differential equation $x''(t) + a(t)x(t) = 0$.

A unifying technique of proof for such estimates can be read in [Mustafa and Rogovchenko (2006)] and is based on the next reformulation of the differential equation (4.27):

$$\begin{cases} y(t) = t^{-\varepsilon} \left[d - \int_t^{+\infty} \tau^\varepsilon f(\tau, x(\tau)) d\tau \right], \\ x(t) = [c - d(\mathrm{sgn}\ \varepsilon - 1)]t + \varepsilon t \int_t^{+\infty} \frac{y(\tau)}{\tau} d\tau - (1 - \varepsilon) \int_{t_0}^t y(\tau) d\tau \end{cases}$$

for some $t_0 \geq 1$ large enough. For a different approach, the so-called Riccatian method, in the case of *intermediate* asymptotic ($\varepsilon \in (0, 1)$, $c = 0$), see the technique from [Agarwal and Mustafa (2007); Mustafa (2007b)].

The study of asymptotically linear solutions to linear and nonlinear ordinary differential equations is of importance in fluid mechanics, differential geometry (Jacobi fields, e.g., [Lang (1999), p. 239]), bidimensional gravity (the geodesics of the euclidean planar spray $x'' = 0$ being the asymptotically linear solutions $x(t) = ct + d$) and others.

Here, we are interested in the existence of a fractional variant for the problem of asymptotically linear solutions which can be formulated as follows: *are there any non-trivial fractional differential equations which have only asymptotically linear solutions and also their solution sets contain solutions (asymptotically linear) for all the prescribed values of numbers c, d and ε?*

To tackle this problem, we shall use a variant of Theorem 4.3. In fact, we shall produce some simple conditions regarding the continuous function $a : \mathbb{R}_+ \to \mathbb{R}$ which ensure that, given $c \in \mathbb{R} - \{0\}$, the fractional differential equation below

$$_0D_t^\alpha (tx' - x + x(0))(t) + a(t)x(t) = 0, \quad t > 0, \tag{4.28}$$

possesses a solution with the asymptotic development $x(t) = c \cdot t + x(0) + o(1)$ when $t \to +\infty$. As before, $tx' - x + x(0)$ designates the mapping $t \mapsto [tx'(t) - x(t) + x(0)] \cdot \chi_{(0,+\infty)}(t)$. The conditions are similar to (4.23).

Let us start with a result regarding the case of intermediate asymptotic.

Proposition 4.3. *Set the numbers* $\varepsilon \in (0, 1)$, $c \neq 0$, *and* $c_1 \in (0, 1)$, $A > 0$ *such that*

$$\max \left\{ |c|, \frac{1}{1 - \varepsilon} \right\} \cdot \Gamma(1 - \alpha)A \leq c_1. \tag{4.29}$$

Assume also that $a \in C([0, +\infty), \mathbb{R})$ is confined to

$$(1 + t^{1-\varepsilon})|a(t)| \le \frac{A}{t^\alpha}, \quad t > 0. \tag{4.30}$$

Then, the fractional differential equation (4.21) has a (mild) solution $x \in C([0, +\infty), \mathbb{R}) \cap C^1((0, +\infty), \mathbb{R})$, with $\lim_{t \searrow 0}[t^{2-\alpha}x'(t)] = 0$, which verifies the asymptotic formula $x(t) = c \cdot t + O(t^\varepsilon)$ when $t \to +\infty$.

Proof. Introduce the complete metric space $\mathcal{M} = (D, \delta)$, where $D = \{y \in C((0, +\infty), \mathbb{R}) : \sup_{t>0}[t^{-\varepsilon}|y(t)|] \le c_1, \, t > 0\}$ and the metric δ is given by the usual formula

$$\delta(y_1, y_2) = \sup_{t>0} \frac{|y_1(t) - y_2(t)|}{t^\varepsilon}, \quad y_1, y_2 \in D.$$

In particular, $\lim_{t \searrow 0} y(t) = 0$ for all $y \in D$.

Recall the function $x : (0, +\infty) \to \mathbb{R}$ displayed in the formulas (1.63). Since $\lim_{t \searrow 0} x(t) = 0$, we deduce that x can be continued backward to 0, so, it has an extension, with the same notation, which belongs to $C([0, +\infty), \mathbb{R}) \cap C^1((0, +\infty), \mathbb{R})$. Also, $\lim_{t \searrow 0}[t^{1-\alpha}y(t)] = \lim_{t \searrow 0}[t^{2-\alpha}x'(t)] = 0$.

Define further the integral operator $\mathcal{T} : \mathcal{M} \to \mathcal{M}$ by the formula

$$(\mathcal{T})(y)(t) = -\frac{1}{\Gamma(\alpha)} \int_0^t \frac{a(s)}{(t-s)^{1-\alpha}} \left[cs - s \int_s^{+\infty} \frac{y(\tau)}{\tau^2} d\tau\right] ds, \quad t > 0.$$

The estimate

$$
\begin{aligned}
&|(\mathcal{T})(y)(t)| \\
&\le \frac{1}{\Gamma(\alpha)} \int_0^t \frac{|a(s)|}{(t-s)^{1-\alpha}} \left(|c|s + \frac{c_1}{1-\varepsilon}s^\varepsilon\right) ds \\
&\le \frac{t^\varepsilon}{\Gamma(\alpha)} \int_0^t \frac{|a(s)|}{(t-s)^{1-\alpha}} \left(\frac{c_1}{1-\varepsilon} + |c|s^{1-\varepsilon}\right) ds \\
&\le \frac{t^\varepsilon}{\Gamma(\alpha)} \int_0^t \frac{|a(s)|}{(t-s)^{1-\alpha}}(1 + s^{1-\varepsilon})ds \cdot \max\left\{|c|, \frac{c_1}{1-\varepsilon}\right\} \quad (4.31) \\
&\le \frac{t^\varepsilon}{\Gamma(\alpha)} \int_0^t \frac{ds}{(t-s)^{1-\alpha}s^\alpha} \cdot A \max\left\{|c|, \frac{c_1}{1-\varepsilon}\right\} \\
&= \Gamma(1-\alpha) \cdot A \max\left\{|c|, \frac{1}{1-\varepsilon}\right\} \cdot t^\varepsilon \\
&\le c_1 t^\varepsilon, \quad t > 0,
\end{aligned}
$$

shows that $\mathcal{T}$ is well-defined by taking into account (4.29), (4.30).

Now, given $y_1, y_2 \in D$, we have

$$|(\mathcal{T})(y_1)(t) - (\mathcal{T})(y_2)(t)|$$

$$\leq \frac{1}{\Gamma(\alpha)} \int_0^t \frac{|a(s)|}{(t-s)^{1-\alpha}} \cdot s \int_s^{+\infty} \frac{d\tau}{\tau^{2-\varepsilon}} ds \cdot \delta(y_1, y_2)$$

$$\leq \frac{1}{\Gamma(\alpha)(1-\varepsilon)} \int_0^t \frac{s^\varepsilon |a(s)|}{(t-s)^{1-\alpha}} ds \cdot \delta(y_1, y_2)$$

$$\leq \frac{1}{\Gamma(\alpha)} \cdot \max\left\{ |c|, \frac{1}{1-\varepsilon} \right\} \cdot \int_0^t \frac{|a(s)|}{(t-s)^{1-\alpha}} ds \cdot t^\varepsilon \cdot \delta(y_1, y_2)$$

$$\leq \frac{1}{\Gamma(\alpha)} \max\left\{ |c|, \frac{1}{1-\varepsilon} \right\} \cdot \int_0^t \frac{(1+s^{1-\varepsilon})|a(s)|}{(t-s)^{1-\alpha}} ds \cdot t^\varepsilon \delta(y_1, y_2)$$

$$\leq \Gamma(1-\alpha) \cdot A \max\left\{ |c|, \frac{1}{1-\varepsilon} \right\} \cdot t^\varepsilon \delta(y_1, y_2)$$

$$\leq t^\varepsilon \cdot c_1 \delta(y_1, y_2), \quad t > 0,$$

and so $\delta(T(y_1), T(y_2)) \leq c_1 \delta(y_1, y_2)$.

The operator $\mathcal{T}$ being a contraction, it has a unique fixed point $y_0 \in D$. Since $t \int_t^{+\infty} \frac{y_0(s)}{s^2} ds = O(t^\varepsilon)$ when $t \to +\infty$, the proof is complete. $\square$

Theorem 4.4. *Assume that (4.29) holds true and $a \in C(\mathbb{R}_+, \mathbb{R})$ verifies the sharper restriction*

$$(1 + t^{1-\varepsilon})|a(t)| \leq A \min\left\{ \frac{1}{t^\alpha}, \frac{1}{t^\beta} \right\}, \quad t > 0, \tag{4.32}$$

where $1 > \beta > \alpha + \varepsilon$. Then, the solution x of (4.21) from Proposition 4.3 has the asymptotic development $x(t) = c \cdot t + o(1)$ when $t \to +\infty$.

Proof. Notice that

$$\int_0^t \frac{ds}{(t-s)^{1-\alpha} s^\beta} = t^{\alpha-\beta} \int_0^1 \frac{du}{(1-u)^{1-\alpha} u^\beta} = t^{\alpha-\beta} B(\alpha, 1-\beta).$$

Via (4.32) and (4.31), we have the estimate

$$|y_0(t)| = |\mathcal{T}(y_0)(t)|$$

$$\leq \frac{t^\varepsilon}{\Gamma(\alpha)} \int_0^t \frac{(1+s^{1-\varepsilon})|a(s)|}{(t-s)^{1-\alpha}} ds \cdot \max\left\{ |c|, \frac{c_1}{1-\varepsilon} \right\}$$

$$\leq \frac{t^\varepsilon}{\Gamma(\alpha)} \int_0^t \frac{ds}{(t-s)^{1-\alpha} s^\beta} \cdot A \max\left\{ |c|, \frac{c_1}{1-\varepsilon} \right\}$$

$$= t^{\varepsilon+\alpha-\beta} \cdot \frac{\Gamma(1-\beta)}{\Gamma(\alpha+1-\beta)} A \max\left\{ |c|, \frac{1}{1-\varepsilon} \right\}$$

$$= o(1) \quad \text{when } t \to +\infty.$$

By means of L'Hôpital's rule, we conclude that — recall (1.63) —

$$\lim_{t \to +\infty} \left[t \int_t^{+\infty} \frac{y_0(s)}{s^2} ds \right] = \lim_{t \to +\infty} y_0(t) = 0.$$

The proof is complete. $\square$

Our main contribution in this section is given next.

Theorem 4.5. ([Baleanu et al. (2010c), Theorem 2.3]) *Set the numbers* $\varepsilon \in (0, 1 - \alpha)$, $\beta \in (\alpha + \varepsilon, 1)$, c, d *with* $c^2 + d^2 > 0$, *and* $c_1 \in (0, 1)$, $A > 0$, *such that*

$$\max \left\{ |c|, |d|, \frac{1}{1 - \varepsilon} \right\} \cdot \Gamma(1 - \alpha) A \le c_1. \tag{4.33}$$

Assume also that $a \in C(\mathbb{R}_+, \mathbb{R})$ *satisfies the inequality*

$$\left(\frac{1}{t^\varepsilon} + 1 + t^{1-\varepsilon} \right) |a(t)| \le A \min \left\{ \frac{1}{t^\alpha}, \frac{1}{t^\beta} \right\}, \quad t > 0. \tag{4.34}$$

Then, the fractional differential equation (4.28) has a solution $x \in C(\mathbb{R}_+, \mathbb{R}) \cap C^1((0, +\infty), \mathbb{R})$, *with* $x(0) = d$ *and* $\lim_{t \searrow 0}[t^{2-\alpha} x'(t)] = 0$, *which has the asymptotic development*

$$x(t) = c \cdot t + d + o(1) \quad \text{when } t \to +\infty.$$

Proof. Keeping the notations from Proposition 4.3, introduce the change of variables

$$y = tx' - x + d, \quad x(t) = ct + d - t \int_t^{+\infty} \frac{y(s)}{s^2} ds, \quad t > 0, \, y \in D,$$

and the integral operator $T : \mathcal{M} \to \mathcal{M}$ with the formula

$$(T)(y)(t) = -\frac{1}{\Gamma(\alpha)} \int_0^t \frac{a(s)}{(t-s)^{1-\alpha}} \left[cs + d - s \int_s^{+\infty} \frac{y(\tau)}{\tau^2} d\tau \right] ds, \quad t > 0.$$

As before, we have the estimates

$$|T(y)(t)|$$
$$\le t^\varepsilon \frac{1}{\Gamma(\alpha)} \int_0^t \frac{|a(s)|}{(t-s)^{1-\alpha}} \left(\frac{1}{s^\varepsilon} + 1 + s^{1-\varepsilon} \right) ds \cdot \max \left\{ |c|, |d|, \frac{c_1}{1-\varepsilon} \right\}$$
$$\le c_1 t^\varepsilon, \quad t > 0,$$

and

$$|(T)(y_1)(t) - (T)(y_2)(t)|$$
$$\le \frac{t^\varepsilon}{\Gamma(\alpha)} \max \left\{ |c|, |d|, \frac{1}{1-\varepsilon} \right\} \int_0^t \frac{\left(\frac{1}{s^\varepsilon} + 1 + s^{1-\varepsilon} \right) |a(s)|}{(t-s)^{1-\alpha}} ds \cdot \delta(y_1, y_2)$$
$$\le t^\varepsilon \cdot c_1 \delta(y_1, y_2), \quad t > 0,$$

for all $y, y_1, y_2 \in D$.

Finally, for the fixed point y_0 of the operator $\mathcal{T}$, we have that

$$|y_0(t)| = |\mathcal{T}(y_0)(t)|$$
$$\leq t^{\varepsilon + \alpha - \beta} \cdot \frac{\Gamma(1 - \beta)}{\Gamma(\alpha + 1 - \beta)} A \max \left\{ |c|, |d|, \frac{1}{1 - \varepsilon} \right\}$$
$$= o(1) \quad \text{when } t \to +\infty.$$

The proof is complete. $\square$

4.3   A Bihari-Like Result

Consider the fractional differential equation

$$ {}_0^1\mathcal{O}_t^{1+\alpha}(x)(t) + f(t, x(t)) = 0, \quad t > 0, \tag{4.35}$$

where the continuous function $f : [0, +\infty) \times \mathbb{R} \to \mathbb{R}$ satisfies the restriction

$$ |f(t, x)| \leq F\left(t, \frac{|x|}{(1 + t)^\alpha}\right), \quad t \geq 0, x \in \mathbb{R}, \tag{4.36}$$

for some continuous comparison function $F : \mathbb{R}_+^2 \to \mathbb{R}_+$ which is assumed nondecreasing in the second argument.

We shall focus here on answering to the following question: *given a solution x of the linear fractional differential equation ${}_0^1\mathcal{O}_t^{1+\alpha}(x)(t) = 0$, is there a solution y of its perturbation (4.35) which deviates "slightly" from x?*

To describe the "noise" f, we opt for the power-like behavior, as in the next fractional Emden-Fowler type of equation

$$ {}_0^1\mathcal{O}_t^{1+\alpha}(x)(t) + a(t)[x(t)]^\lambda = 0, \quad t > 0, \tag{4.37}$$

where $a(t) \sim t^\mu$ when $t \to +\infty$ for some $\mu < 0$, $\lambda > 1$ and $x^\lambda = |x|^{\lambda-1}x$. We recall that this equation has been used, in its classical counterpart, to model gaseous dynamics in astrophysics. Since the solutions of the linear part are 1 and t^α, comparing them with the solutions of (4.37) means rawly a comparison with $\max\{1, t^\alpha\}$. Such an estimate of the potential *growth in time* for the solutions of (4.35) have lead us to the restriction (4.36), of Bihari type. The restriction has been used in the case of classical differential equations — that is, with $\alpha = 1$ — to deal with various variants of (4.37), see the comprehensive monograph [Bellman (1953)]. For a survey of the literature on the second order ordinary differential equations with Bihari-like nonlinearity, the reader can consult [Mustafa and Rogovchenko (2002); Agarwal et al. (2007a)].

Set the numbers a, $b \in \mathbb{R}$, with $a^2 + b^2 > 0$, and

$$\varepsilon \in \left(0, \frac{1}{2}[|b| + \operatorname{sgn}(1 - \operatorname{sgn}|b|)] \right).$$

Assume also that $\int_1^{+\infty} tF(t, |a| + |b| + \varepsilon)dt < +\infty$ and

$$\int_0^{+\infty} F(t, |a| + |b| + \varepsilon)dt \le \varepsilon\Gamma(1 + \alpha). \tag{4.38}$$

Theorem 4.6. ([Baleanu et al. (2011a), Theorem 1]) *The fractional differential equation (4.35) has a (strong) solution* $x \in C([0, +\infty), \mathbb{R}) \cap C^1((0, +\infty), \mathbb{R})$, *with* $\lim_{t \searrow 0}[t^{1-\alpha}x'(t)] \in \mathbb{R}$, *such that*

$$x(t) = a + bt^\alpha + O(t^{\alpha-1}) = a + bt^\alpha + o(1) \quad \text{when } t \to +\infty. \tag{4.39}$$

Proof. Introduce the real linear space X of all the functions $x \in C(\mathbb{R}_+, \mathbb{R})$ such that $\frac{|x(t)|}{(1+t)^\alpha} \le x_{\infty,\alpha} < +\infty$, where $t \ge 0$, endowed with the usual operations with numeric functions. A natural norm on X reads as

$$\|x\| = \sup_{t \ge 0} \frac{|x(t)|}{(1+t)^\alpha},$$

making $\mathcal{X} = (X, \| \star \|)$ a Banach function space.

Consider further the convex set

$$D = \{x \in X : |x(t) - a - bt^\alpha| \le \varepsilon(1 + t)^\alpha, \ t \ge 0\}$$

and introduce the integral operator $\mathcal{T} : D \to C(\mathbb{R}_+, \mathbb{R})$ by means of formula

$$\mathcal{T}(x)(t) = a + bt^\alpha + \frac{1}{\Gamma(\alpha)} \int_0^t \frac{1}{(t-s)^{1-\alpha}} \int_s^{+\infty} f(\tau, x(\tau))d\tau ds, \quad t > 0.$$

Given $x \in D$, we have

$$|x(t)| \le |a| + |b|t^\alpha + \varepsilon(1+t)^\alpha \le (|a| + |b| + \varepsilon)(1+t)^\alpha \tag{4.40}$$

and, via (4.38),

$$\begin{aligned}
|\mathcal{T}(x)(t) - a - bt^\alpha| &\le \frac{1}{\Gamma(\alpha)} \int_0^t \frac{1}{(t-s)^{1-\alpha}} \int_s^{+\infty} F\left(\tau, \frac{|x(\tau)|}{(1+\tau)^\alpha}\right) d\tau ds \\
&\le \frac{1}{\Gamma(\alpha)} \int_0^t \frac{ds}{(t-s)^{1-\alpha}} \cdot \int_0^{+\infty} F(\tau, |a| + |b| + \varepsilon)d\tau \\
&= \frac{t^\alpha}{\Gamma(1+\alpha)} \int_0^{+\infty} F(\tau, |a| + |b| + \varepsilon)d\tau \\
&\le \varepsilon(1+t)^\alpha, \quad t \ge 0,
\end{aligned}$$

which means that $\mathcal{T}(D) \subseteq D$.

We shall prove now that *the set $T(D)$ is relatively compact in $\mathcal{X}$.* To this end, following the technique from [Mustafa and Rogovchenko (2002); Avramescu (1969)], we have to establish that: (i) the set $T(D)$ is bounded; (ii) given $t_2 \geq t_1 \geq 0$, one has

$$\left| \frac{T(x)(t_2)}{(1 + t_2)^\alpha} - \frac{T(x)(t_1)}{(1 + t_1)^\alpha} \right| \leq \chi \cdot \max\left\{ (t_2 - t_1)^\alpha, t_2 - t_1 \right\}, \quad x \in D,$$

where $\chi = \alpha|a| + |b| + 2\varepsilon$; and (iii)

$$\lim_{t \to +\infty} \frac{T(x)(t)}{(1 + t)^\alpha} = b \quad \text{uniformly with respect to } x \in D.$$

The estimate (i) follows from (4.40) and the fact that $T(D) \subseteq D$, namely $\|T(x)\| \leq |a| + |b| + \varepsilon$, $x \in D$.

To prove (ii), start by noticing the elementary estimates

$$\left| \frac{1}{(1 + t_2)^\alpha} - \frac{1}{(1 + t_1)^\alpha} \right| \leq \alpha(t_2 - t_1)$$

and

$$\left| \left(\frac{t_2}{1 + t_2} \right)^\alpha - \left(\frac{t_1}{1 + t_1} \right)^\alpha \right| \leq (t_2 - t_1)^\alpha, \quad t_2 \geq t_1 \geq 0.$$

To deal with the first one, use the mean value theorem, that is

$$\left| \frac{1}{(1 + t_2)^\alpha} - \frac{1}{(1 + t_1)^\alpha} \right| = \frac{\alpha}{(1 + \xi)^{1+\alpha}} \cdot (t_2 - t_1)$$
$$\leq \alpha(t_2 - t_1),$$

where $\xi \in [t_1, t_2]$. As for the second inequality, via (1.8),

$$\left| \left(\frac{t_2}{1 + t_2} \right)^\alpha - \left(\frac{t_1}{1 + t_1} \right)^\alpha \right|$$
$$\leq \left| \frac{t_2}{1 + t_2} - \frac{t_1}{1 + t_1} \right|^\alpha = \frac{(t_2 - t_1)^\alpha}{[(1 + t_1)(1 + t_2)]^\alpha}$$
$$\leq (t_2 - t_1)^\alpha.$$

Further, we have

$$T(x)(t) = a + bt^\alpha + \frac{1}{\Gamma(1 + \alpha)}$$
$$\times \left\{ \int_0^t [t^\alpha - (t - \tau)^\alpha] f(\tau, x(\tau)) d\tau + t^\alpha \int_t^{+\infty} f(\tau, x(\tau)) d\tau \right\} \quad (4.41)$$
$$= a + c(x)t^\alpha - \frac{1}{\Gamma(1 + \alpha)} \int_0^t (t - \tau)^\alpha f(\tau, x(\tau)) d\tau, \quad (4.42)$$

see (4.10), where $c(x) = b + \frac{1}{\Gamma(1+\alpha)} \int_0^{+\infty} f(\tau, x(\tau)) d\tau$, which leads to

$$\left| \frac{\mathcal{T}(x)(t_2)}{(1+t_2)^\alpha} - \frac{\mathcal{T}(x)(t_1)}{(1+t_1)^\alpha} \right|$$

$$\leq |a| \left| \frac{1}{(1+t_2)^\alpha} - \frac{1}{(1+t_1)^\alpha} \right| + |c(x)| \left| \left(\frac{t_2}{1+t_2} \right)^\alpha - \left(\frac{t_1}{1+t_1} \right)^\alpha \right|$$

$$+ \frac{1}{\Gamma(1+\alpha)} \int_0^{t_1} [(t_2 - \tau)^\alpha - (t_1 - \tau)^\alpha] |f(\tau, x(\tau))| d\tau$$

$$+ \frac{1}{\Gamma(1+\alpha)} \int_{t_1}^{t_2} (t_2 - \tau)^\alpha |f(\tau, x(\tau))| d\tau.$$

Since $|c(x)| \leq |b| + \frac{1}{\Gamma(1+\alpha)} \int_0^{+\infty} F(\tau, |a| + |b| + \varepsilon) d\tau \leq |b| + \varepsilon$, we get that

$$\left| \frac{\mathcal{T}(x)(t_2)}{(1+t_2)^\alpha} - \frac{\mathcal{T}(x)(t_1)}{(1+t_1)^\alpha} \right|$$

$$\leq |a|\alpha(t_2 - t_1) + (|b| + \varepsilon)(t_2 - t_1)^\alpha$$

$$+ \frac{1}{\Gamma(1+\alpha)} \int_0^{t_1} (t_2 - t_1)^\alpha F(\tau, |a| + |b| + \varepsilon) d\tau$$

$$+ \frac{1}{\Gamma(1+\alpha)} \int_{t_1}^{t_2} (t_2 - t_1)^\alpha F(\tau, |a| + |b| + \varepsilon) d\tau$$

$$\leq |a|\alpha(t_2 - t_1) + (|b| + \varepsilon)(t_2 - t_1)^\alpha$$

$$+ \frac{1}{\Gamma(1+\alpha)} \int_0^{+\infty} F(\tau, |a| + |b| + \varepsilon) d\tau \cdot (t_2 - t_1)^\alpha$$

$$\leq (\alpha|a| + |b| + 2\varepsilon) \cdot \max\left\{ (t_2 - t_1)^\alpha, t_2 - t_1 \right\}.$$

To prove (iii), recalling (4.41), we rely on the estimates

$$\left| \int_0^t [t^\alpha - (t - \tau)^\alpha] f(\tau, x(\tau)) d\tau + t^\alpha \int_t^{+\infty} f(\tau, x(\tau)) d\tau \right|$$

$$\leq t^\alpha \left[\int_0^t \frac{\tau}{t} |f(\tau, x(\tau))| d\tau + \int_t^{+\infty} |f(\tau, x(\tau))| d\tau \right]$$

$$\leq t^{\alpha-1} \int_0^{+\infty} \tau F(\tau, |a| + |b| + \varepsilon) d\tau$$

$$= O(t^{\alpha-1}) \quad \text{when } t \to +\infty, \tag{4.43}$$

which means that

$$\left| \frac{\mathcal{T}(x)(t)}{(1+t)^\alpha} - b \right| \leq \frac{|\mathcal{T}(x)(t) - a - bt^\alpha|}{(1+t)^\alpha} + \frac{|a|}{(1+t)^\alpha} + |b| \left| 1 - \left(\frac{t}{1+t} \right)^\alpha \right|$$

$$\leq \frac{|a|}{(1+t)^\alpha} + |b| \left| 1 - \left(\frac{t}{1+t} \right)^\alpha \right|$$

$$+ \frac{1}{t} \cdot \left(\frac{t}{1+t} \right)^\alpha \int_0^{+\infty} \tau F(\tau, |a| + |b| + \varepsilon) d\tau$$

$$= o(1) \quad \text{when } t \to +\infty.$$

The relative compactness of $T(D)$ in $\mathcal{X}$ is now established. As a by-product, if $x \in D$, we get $T(x)(t) = a + bt^\alpha + O(t^{\alpha-1}) = a + [b + o(1)]t^\alpha$ for all the large values of t.

To prove that *the operator* $T : D \to D$ *is continuous*, set $\delta > 0$ and notice the uniform continuity of $f(t, x)$ in $[0, T] \times [-A, A]$, where $T > 0$ is large enough to have $\int_T^{+\infty} F(t, |a| + |b| + \varepsilon) dt < \frac{\delta}{8}$ and $A = A(T) = (|a| + |b| + \varepsilon)(1 + T)^\alpha$. This yields

$$|f(t, x) - f(t, y)| \leq \frac{\delta}{4T}, \quad t \in [0, T],$$

for all x, $y \in [-A, A]$ with $|x - y| \leq \xi = \xi(\delta, T)$ small enough.

We have the estimates

$$|T(x)(t) - T(y)(t)|$$

$$\leq |c(x) - c(y)|t^\alpha + \frac{1}{\Gamma(1+\alpha)} \int_0^t (t - \tau)^\alpha |f(\tau, x(\tau)) - f(\tau, y(\tau))| d\tau$$

$$\leq t^\alpha \cdot \frac{1}{\Gamma(1+\alpha)} \int_0^{+\infty} |f(\tau, x(\tau)) - f(\tau, y(\tau))| d\tau$$

$$+ \frac{1}{\Gamma(1+\alpha)} \int_0^t (t - \tau)^\alpha |f(\tau, x(\tau)) - f(\tau, y(\tau))| d\tau$$

$$\leq \frac{2t^\alpha}{\Gamma(1+\alpha)} \left[\int_0^T |f(\tau, x(\tau)) - f(\tau, y(\tau))| d\tau \right.$$

$$+ \left. \int_T^{+\infty} (|f(\tau, x(\tau))| + |f(\tau, y(\tau))|) d\tau \right]$$

$$\leq \frac{2t^\alpha}{\Gamma(1+\alpha)} \left[\frac{\delta}{4T} \cdot T + 2 \int_T^{+\infty} F(\tau, |a| + |b| + \varepsilon) d\tau \right]$$

$$\leq \frac{\delta}{\Gamma(1+\alpha)} \cdot (1+t)^\alpha, \quad t \geq 0,$$

and thus $\|T(x) - T(y)\| \leq \frac{\delta}{\Gamma(1+\alpha)}$ for all x, $y \in D$ with $\|x - y\| \leq \xi$.

Now, since the operator T is compact (completely continuous), the Schauder fixed point theorem yields the existence of at least one fixed point of T, denoted with u, in D. By taking into account (4.43), we deduce that $u(t) = T(u)(t) = a + bt^\alpha + O(t^{\alpha-1})$ when $t \to +\infty$.

The last step of the proof is to establish that $u \in C^1((0, +\infty), \mathbb{R})$ with $\lim_{t \searrow 0}[t^{1-\alpha}u'(t)] \in \mathbb{R}$. Let us recall that u can be expressed as in (4.42), which means we only have to prove that *the function* $t \mapsto \mathcal{I}(t) = \int_0^t (t - \tau)^\alpha f(\tau, u(\tau))d\tau$ *is continuously differentiable in* $(0, +\infty)$. The differentiability of $\mathcal{I}$ everywhere follows from (1.20) while the continuity of $\mathcal{I}' = Q_{1-\alpha,\alpha}f(t, u)$ is a consequence of (1.45). Here, by $f(t, u)$ we understand the mapping $t \mapsto f(t, u(t))$.

We get

$$u'(t) = \alpha c(u)t^{\alpha-1} - \frac{1}{\Gamma(\alpha)} \int_0^t \frac{f(\tau, u(\tau))}{(t - \tau)^{1-\alpha}}d\tau, \quad t > 0. \qquad (4.44)$$

As $f(t, u) \in C([0, +\infty), \mathbb{R})$, we deduce that $\lim_{t \searrow 0}[t^{1-\alpha}u'(t)] = \alpha c(u)$ via (1.34).

By an integration of (4.44), we obtain that

$$\int_0^t \frac{u'(s)}{(t - s)^\alpha}ds$$

$$= \alpha c(u) \cdot B(\alpha, 1 - \alpha) - \frac{1}{\Gamma(\alpha)} \int_0^t \frac{1}{(t - s)^\alpha} \int_0^s \frac{f(\tau, u(\tau))}{(s - \tau)^{1-\alpha}}d\tau ds$$

$$= \alpha c(u)B(\alpha, 1 - \alpha) - \frac{1}{\Gamma(\alpha)} \int_0^t f(\tau, u(\tau)) \int_\tau^t \frac{ds}{(t - s)^\alpha(s - \tau)^{1-\alpha}}d\tau$$

$$= B(\alpha, 1 - \alpha) \left[\alpha c(u) - \frac{1}{\Gamma(\alpha)} \int_0^t f(\tau, u(\tau))d\tau \right], \quad t > 0,$$

which implies *the existence of the quantity* $_0D_t^\alpha(u')$ *throughout* $(0, +\infty)$. Finally, differentiating with respect to the time variable of this integral equality, we arrive at (4.35).

The proof is complete. $\square$

We shall particularize now the restrictions of the theorem in the cases of Emden-Fowler (4.37) and respectively Lipschitz-like fractional differential equations. Firstly, assuming that the coefficient $a : [0, +\infty) \to \mathbb{R}$ is continuous, we have $F(t, u) = (1 + t)^{\alpha\lambda}|a(t)||u|^\lambda$. So, the restrictions look like $\int_1^{+\infty} t^{1+\alpha\lambda}|a(t)|dt < +\infty$ and

$$\int_0^{+\infty} t(1 + t)^{\alpha\lambda}|a(t)|dt \leq \frac{\varepsilon}{(|a| + |b| + \varepsilon)^\lambda} \cdot \Gamma(1 + \alpha).$$

Secondly, suppose that the nonlinearity f from (4.35) obeys a Lipschitz-like condition

$$|f(t, x) - f(t, y)| \leq h(t)|x - y|, \quad t \geq 0, \, x, y \in \mathbb{R},$$

for some continuous function $h : \mathbb{R}_+ \to \mathbb{R}_+$. Then, the nonlinearity f will verify also the inequality (4.36) for $F(t, u) = (1 + t)^\alpha h(t)|u| + |f(t, 0)|$. In this case, the hypotheses of our theorem read as $\int_1^{+\infty} t^{1+\alpha} h(t)dt < +\infty$, $\int_1^{+\infty} t|f(t, 0)|dt < +\infty$ and

$$(|a| + |b| + \varepsilon) \int_0^{+\infty} (1 + t)^\alpha h(t)dt + \int_0^{+\infty} |f(t, 0)|dt \leq \varepsilon \Gamma(1 + \alpha).$$

Let us conclude this section with an illustrative example concerning the error bound in the development (4.39). Ideally, one can establish that an error is undiminishable by presenting a carefully designed example, as was done e.g., in [Baleanu and Mustafa (2009)]. It is a known fact, however, that most of the fractional differential equations do not possess solutions expressible in closed form, meaning we have to deal in most circumstances with complicated power series and to lean upon deep theorems from the special functions theory to get some decent estimate of the sums of these power series. The same setback is encountered in the (classical) asymptotic integration theory of ordinary differential equations, see [Mustafa and Rogovchenko (2002)]. In our case, fortunately, there is an elementary way out leading to some "borderline situation".

Notice first that the main restriction (4.36) subsumes the case of *perturbed* Emden-Fowler fractional differential equations, that is

$$\prescript{1}{0}{\mathcal{O}}_t^{1+\alpha}(x)(t) + q(t)[x(t)]^\lambda = p(t), \quad t > 0, \tag{4.45}$$

where the coefficient q and the perturbing term $p : \mathbb{R}_+ \to \mathbb{R}$ are continuous functions. Here, the comparison function F reads as $F(t, u) = (1 + t)^{\alpha\lambda}|q(t)|u^\lambda + |p(t)|$.

Now, to deal with the error bound $O(t^{\alpha-1})$ when $t \to +\infty$, let us see what happens if we take $x(t) = -\int_t^{+\infty} A(s)ds, t \geq 0$, where $A : [0, +\infty) \to \mathbb{R}$ is some C^1–function with $A(t) = O(t^{\alpha-2})$ when $t \to +\infty$. Meaning, we have to see *which* are the functions q, p from (4.45) and *if* the comparison function F given previously obeys the hypotheses of the theorem.

We shall use the technique from [Baleanu et al. (2010d), Section 3]. To this end, assume that the function A has a unique zero $T > 0$, with $A(t) > 0$ in $[0, T)$ and $A(t) < 0$ for $t > T$, and $\int_0^{+\infty} A(s)ds = 0$, $\int_0^{+\infty} t|A(t)|dt < +\infty$. Also, $B \in (L^1 \cap L^\infty)((T, +\infty), \mathbb{R})$ for the function B with $B(t) = t^{1-\alpha}\|A|_{[t,+\infty)}\|_{L^\infty((t,+\infty),\mathbb{R})}$. A simple candidate for B is provided by the restriction $|A(t)| \leq c \cdot t^{-(2+\varepsilon)}$, where $t \geq T$, for some $c, \varepsilon > 0$.

In particular, while being negative valued in $(T, +\infty)$, the function A increases eventually to 0 when $t \to +\infty$, as suggested by the existence of

B. So, it is natural to ask for a global minimum of A, attained in $T_1 > T$. We impose as well that $A'(t) < 0$ in $[0, T_1)$ and $A'(t) > 0$ for $t > T_1$.

Observe that, since[5] $\lim_{t \to +\infty} tA(t) = 0$, an integration by parts yields $\int_0^{+\infty} tA'(t)dt = 0$. The last restrictions are given by $C \in (L^1 \cap L^\infty)((T_1, +\infty), \mathbb{R})$, where $C(t) = t^{1-\alpha} \| [tA'(t)] |_{[t,+\infty)} \|_{L^\infty((t,+\infty),\mathbb{R})}$ and[6] $\int_0^{+\infty} t^2 |A'(t)| dt < +\infty$.

In other words, we have asked that *the functions A and $t \mapsto tA'(t)$ be both members of the class of functions described in [Baleanu et al. (2010d)]*. A necessary coupling was needed, provided by the condition $T_1 > T$.

Now, *we claim that*

$$\int_0^{+\infty} \left| \int_0^t \frac{A(s)}{(t-s)^\alpha} ds \right| dt + \sup_{t \ge 0} \left| \int_0^t \frac{A(s)}{(t-s)^\alpha} ds \right| < +\infty. \qquad (4.46)$$

To demonstrate the validity of this statement, start with the computation

$$\left| \int_0^{2t} \frac{A(s)}{(2t-s)^\alpha} ds \right| \le \left| \int_0^t \frac{A(s)}{(2t-s)^\alpha} ds \right| + \int_t^{2t} \frac{ds}{(2t-s)^\alpha} \cdot t^{\alpha-1} B(t)$$

$$= \left| \int_0^t \frac{A(s)}{(2t-s)^\alpha} ds \right| + \frac{B(t)}{1-\alpha}, \quad t \ge T.$$

Further,

$$\int_0^t \frac{A(s)}{(2t-s)^\alpha} ds = \int_0^t \frac{1}{(2t-s)^\alpha} \cdot \left(\int_0^s A(\tau) d\tau \right)' ds$$

$$= \frac{1}{t^\alpha} \int_0^t A(\tau) d\tau - \alpha \int_0^t \frac{1}{(2t-s)^{1+\alpha}} \left(\int_0^s A(\tau) d\tau \right) ds \qquad (4.47)$$

$$= -\alpha \int_0^t \frac{1}{(2t-s)^{1+\alpha}} \left(\int_0^s A(\tau) d\tau \right) ds + o(t^{-\alpha}) \quad \text{when } t \to +\infty,$$

since $\int_0^{+\infty} A(s)ds = \lim_{t \to +\infty} \int_0^t A(s)ds = 0$.

Now, we have

$$\left| \int_0^t \frac{1}{(2t-s)^{1+\alpha}} \left(\int_0^s A(\tau) d\tau \right) ds \right| \le \int_0^t \frac{1}{(2t-s)^{1+\alpha}} \left| \int_0^s A(\tau) d\tau \right| ds$$

---

[5] We have $\int_0^{+\infty} |A(t)| dt < +\infty$ and, given eventual monotonicity of A, a verbatim application of the technique (3.26) will lead to the existence of this limit. We have requested as well that $A(t) = O(t^{\alpha-2})$ when $t \to +\infty$, which also determines the existence of the limit.

[6] Recall the sign of A'! Another integration by parts shows that, for the integral to exist, it is enough that $\lim_{t \to +\infty} [t^2 A(t)] \in \mathbb{R}$.

$$\leq \int_0^t \frac{1}{t^{1+\alpha}} \left| \int_0^s A(\tau)d\tau \right| ds \leq \frac{1}{t^{1+\alpha}} \int_0^{+\infty} \left| \int_0^s A(\tau)d\tau \right| ds \qquad (4.48)$$

$$= O(t^{-(1+\alpha)}) \quad \text{when } t \to +\infty,$$

since

$$\int_{2T}^{+\infty} \left| \int_0^s A(\tau)d\tau \right| ds = \int_{2T}^{+\infty} \left| \left(\int_0^T + \int_T^s \right) A(\tau)d\tau \right| ds$$

$$= \int_{2T}^{+\infty} \left| - \int_s^{+\infty} A(\tau)d\tau \right| ds = \int_{2T}^{+\infty} \int_s^{+\infty} |A(\tau)|d\tau ds$$

$$= \int_{2T}^{+\infty} (s - 2T)|A(s)|ds < +\infty.$$

We have obtained that $\lim\limits_{t \to +\infty} \int_0^t \frac{A(s)}{(t-s)^\alpha} ds = 0$. Since $\lim\limits_{t \searrow 0} \int_0^t \frac{A(s)}{(t-s)^\alpha} ds = 0$ and the application $t \mapsto \int_0^t \frac{A(s)}{(t-s)^\alpha} ds$ is continuous[7], we deduce that *it is bounded over* $[0, +\infty)$.

Finally, by means of (4.47), (4.48),

$$\int_{4T}^{+\infty} \left| \int_0^{2t} \frac{A(s)}{(2t-s)^\alpha} ds \right| dt$$

$$\leq \frac{1}{1-\alpha} \int_{4T}^{+\infty} B(t)dt + \frac{1}{(4T)^\alpha} \int_{4T}^{+\infty} \left| \int_0^t A(\tau)d\tau \right| dt$$

$$+ \alpha \int_{4T}^{+\infty} \frac{dt}{t^{1+\alpha}} \cdot \int_0^{+\infty} \left| \int_0^s A(\tau)d\tau \right| ds < +\infty.$$

The claim is established. According to our leading assumption, *the mapping* $t \mapsto tA'(t)$ *verifies also (4.46)*.

We have

$$\Gamma(1 - \alpha) \cdot {}_0 O_t^{1+\alpha}(x)(t) = \frac{1-\alpha}{t} \int_0^t \frac{A(s)}{(t-s)^\alpha} ds + \frac{1}{t} \int_0^t \frac{sA'(s)}{(t-s)^\alpha} ds$$

for every $t > 0$, recall (1.11). So, let us take in (4.45) $q(t) = 0$ and $p(t) = \frac{1}{\Gamma(1-\alpha) \cdot t} \int_0^t \frac{1}{(t-s)^\alpha} [(1-\alpha)A(s) + sA'(s)]ds$ for $t > 0$. Although $p \in C((0, +\infty), \mathbb{R})$ is not defined in $t = 0$, we have $\int_0^{+\infty} t|p(t)|dt < +\infty$ thanks to the estimate (4.46), which means that we are "on the border" of the domain of applicability for our theorem.

The trial solution x fits the "profile" $x(t) = O(t^{\alpha-1})$ when $t \to +\infty$. Here, the coefficients from (4.39) are $a = - \int_0^{+\infty} A(s)ds = 0$ and $b = 0$. Remark that x is supposed to be the solution of a *two-point boundary value*

---

[7]This mapping is, in fact, $Q_{\alpha,1}A$, see (1.30). For the latter (null) limit, recall (1.34).

problem on $[0, +\infty)$ for a fractional differential equation. As trial-and-error methods and shooting techniques are virtually inexistent for differential equations of non-integer orders, one might appreciate the accuracy provided by the "blind bullet" $-\int_t^{+\infty} A(s)ds$.

4.4   Convergent Solutions

A particular case of Theorems 4.2, 4.6 is of interest here, namely the one when $b = 1$, $c = 0$ in (4.20) or, equivalently, $a = 1$, $b = 0$ in (4.39). Since they have a finite limit as $t \to +\infty$, such solutions of (4.19), (4.35) may be called *convergent*.

We are interested in seing whether, similarly to the circumstances of the ordinary differential equations [Mustafa and Rogovchenko (2007)], [Agarwal et al. (2007b), Section 7], we can hope or not for the following type of behavior

$$x(t) = 1 + o(1) \quad \text{as } t \to +\infty, \quad x' \in (L^1 \cap L^\infty)((0, +\infty), \mathbb{R}), \quad (4.49)$$

where x is a solution to (4.19).

Judging from the inferences in [Baleanu et al. (2010d)], we observe that, most probably, to get such a result one must look for a *sign-changing* functional coefficient a. This conclusion is provided by the claim that, *given* $a \in C(\mathbb{R}_+, \mathbb{R})$, *the quantity* $F(t) = \int_0^t \frac{|a(s)|}{(t-s)^{1-\alpha}}ds$, $t \geq 0$, *does not belong to* $L^1((0, +\infty), \mathbb{R})$. In fact,

$$\int_T^t F(2s)ds \geq \int_T^t \int_{\frac{T}{2}}^{2s} \frac{|a(\tau)|}{(2s-\tau)^{1-\alpha}}d\tau ds \geq \int_T^t \frac{ds}{\left(2s - \frac{T}{2}\right)^{1-\alpha}} \cdot \int_{\frac{T}{2}}^{2T} |a(\tau)|d\tau$$
$$\to +\infty \quad \text{when } t \to +\infty,$$

where $T > 0$ is chosen large enough for a to be non-trivial in $\left[\frac{T}{2}, 2T\right]$.

We shall discuss in the following the issue of "$x' \in L^1$".

Lemma 4.1. *Assume that* $a \in (C \cap L^\infty)([0, +\infty), \mathbb{R})$ *verifies the hypotheses from [Baleanu et al. (2010d)][8]: it has a unique zero* $t_0 > 0$, $\int_0^{+\infty} a(s)ds = 0$, $\int_0^{+\infty} s|a(s)|ds < +\infty$ *and* $B \in (L^1 \cap L^\infty)((0, +\infty), \mathbb{R})$, *where* $B(t) = t^\alpha \|a\|_{L^\infty((t,+\infty),\mathbb{R})}$ *for all* $t \geq 0$. *Then, introducing the quantity* $C(t) = \int_0^t \frac{a(s)}{(t-s)^{1-\alpha}}ds$, $t \geq 0$, *we have*

$$\int_0^{+\infty} |C(t)|dt + \sup_{t \geq 0} |C(t)| < +\infty. \quad (4.50)$$

---

[8]Recall the discussion regarding the class of functions A at page 101 in which α is replaced with $1 - \alpha$.

If $B^* \in L^1((0,+\infty), \mathbb{R})$, *where* $B^*(t) = \|B\|_{L^\infty((t,+\infty),\mathbb{R})}$ *for all* $t \geq 0$, *then*

$$\int_0^{+\infty} \|C\|_{L^\infty((t,+\infty),\mathbb{R})} dt < +\infty, \quad t \geq 0. \qquad (4.51)$$

If $\alpha \in (0, \frac{1}{2})$ *and* $B^{**} \in L^1((0,+\infty), \mathbb{R})$, *where* $B^{**}(t) = \|B\|_{L^2((t,+\infty),\mathbb{R})}$ *for all* $t \geq 0$, *then we also have*

$$\int_0^{+\infty} \|C\|_{L^2((u,+\infty),\mathbb{R})} du < +\infty. \qquad (4.52)$$

Proof. As in [*Baleanu et al. (2010d)*], for $t > 0$, the following estimates are valid

$$|C(2t)| \leq \frac{B(t)}{\alpha} + \left| \int_0^t \frac{a(s)}{(2t-s)^{1-\alpha}} ds \right| \qquad (4.53)$$

and

$$\int_0^t \frac{a(s)}{(2t-s)^{1-\alpha}} ds$$

$$= t^{\alpha-1} \int_0^t a(s) ds - (1-\alpha) \int_0^t \frac{1}{(2t-s)^{2-\alpha}} \int_0^s a(\tau) d\tau ds \qquad (4.54)$$

$$= -t^{\alpha-1} \int_t^{+\infty} a(s) ds$$

$$+ (1-\alpha) \int_0^t \frac{1}{(2t-s)^{2-\alpha}} \int_s^{+\infty} a(\tau) d\tau ds. \qquad (4.55)$$

Since $B \in L^1 \cap L^\infty$, it is obvious that $B \in L^2$, so we shall focus on the second member from the right part of (4.53). By means of (4.54), we get

$$\left| t^{\alpha-1} \int_0^t a(s) ds \right| + \left| \int_0^t \frac{1}{(2t-s)^{2-\alpha}} \int_0^s a(\tau) d\tau ds \right|$$

$$\leq t^\alpha \|a\|_{L^\infty((0,+\infty),\mathbb{R})} + \frac{1}{t^{2-\alpha}} \int_0^t (s \cdot \|a\|_{L^\infty(0,+\infty),\mathbb{R})}) ds$$

$$= \frac{3}{2} \|a\|_{L^\infty(0,+\infty),\mathbb{R})} \cdot t^\alpha,$$

which leads to[9] $C \in (L^1 \cap L^\infty)((0, T_0), \mathbb{R})$, where $T_0 = \max\{1, t_0\}$. Further, via (4.55),

$$D(t) = \left| \int_0^t \frac{a(s)}{(2t-s)^{1-\alpha}} ds \right|$$

$$\leq t^{\alpha-1} \int_t^{+\infty} |a(s)| ds + t^{\alpha-2} \int_0^t \int_s^{+\infty} |a(\tau)| d\tau ds$$

---

[9]In fact, $C \in C([0, T_0], \mathbb{R})$, see the footnote on page 103.

$$= 2t^{\alpha-1} \int_t^{+\infty} |a(s)|ds + t^{\alpha-2} \int_0^t \tau |a(\tau)|d\tau$$

$$\leq \frac{3}{t^{2-\alpha}} \int_0^{+\infty} s|a(s)|ds = \frac{\zeta_a}{t^{2-\alpha}}, \quad t \geq T_0,$$

and so $C \in (L^1 \cap L^\infty)((T_0, +\infty), \mathbb{R})$. The estimate (4.50) has been obtained. As a by-product, $C \in L^2((0, +\infty), \mathbb{R})$.

To prove (4.51), introduce $D^*(t) = \sup_{s \geq t} D(s)$ for all $t \geq 0$. We have that

$$D^*(t) \leq \operatorname{sgn}(T_0 - t)^+ \cdot \frac{3}{2} \|a\|_{L^\infty(0,+\infty),\mathbb{R}} T_0^\alpha$$

$$+ \operatorname{sgn}(t - T_0)^+ \cdot \frac{\zeta_a}{t^{2-\alpha}}, \quad t \geq 0, \tag{4.56}$$

and so $D^* \in (L^1 \cap L^\infty)((0, +\infty), \mathbb{R})$. The conclusion follows from (4.53), since

$$\|C\|_{L^\infty((2t,+\infty),\mathbb{R})} \leq \frac{B^*(t)}{\alpha} + D^*(t), \quad t \geq 0.$$

For the third part, notice that, by means of the triangle inequality, we get

$$\|C\|_{L^2((2t,+\infty),\mathbb{R})} \leq \frac{B^{**}(t)}{\alpha} + D^{**}(t), \quad t \geq 0,$$

where $D^{**}(t) = \|D\|_{L^2((t,+\infty),\mathbb{R})}$. The restriction $\alpha \in \left(0, \frac{1}{2}\right)$ is needed in (4.56) to ensure that $D^{**} \in L^1$.

The proof is complete. $\square$

Lemma 4.2. ([Baleanu et al. (2011b), Lemma 2]) *Assume that the function C from Lemma 4.1 satisfies the restrictions (4.50), (4.51) and (4.52) and either*

$$k_1 = \|C\|_{L^\infty((0,+\infty),\mathbb{R})} + 2\|C^*\|_{L^1((0,+\infty),\mathbb{R})} < 1,$$

where $C^*(t) = \|C\|_{L^\infty((t,+\infty),\mathbb{R})}$, or

$$2\|C^*\|_{L^1((0,+\infty),\mathbb{R})} < 1$$

and

$$k_2 = \max \left\{ \|C\|_{L^\infty((0,+\infty),\mathbb{R})} + \|C\|_{L^2((0,+\infty),\mathbb{R})}, \right.$$

$$\left. \|C\|_{L^1((0,+\infty),\mathbb{R})} + \|E\|_{L^1((0,+\infty),\mathbb{R})} \right\} < 1,$$

where $E(t) = \|C\|_{L^2((t,+\infty),\mathbb{R})}$ for all $t \geq 0$. Then, there exists a function $y \in (C \cap L^1 \cap L^\infty)([0, +\infty), \mathbb{R})$ such that

$$y(t) = -C(t)\left(1 - \int_t^{+\infty} y(s)ds\right) - \int_t^{+\infty} (Cy)(s)ds, \quad t \geq 0. \tag{4.57}$$

Proof. Set the number $\gamma > 1$ such that

$$1 + 2\gamma \int_0^{+\infty} C^*(s)ds < \gamma.$$

Introduce the set Y of all the functions $y \in C([0, +\infty), \mathbb{R})$ such that $|y(t)| \leq \gamma \cdot C^*(t)$, $t \geq 0$, and the metric d with the formula

$$d(y_1, y_2) = \max\{\|y_1 - y_2\|_{L^\infty((0,+\infty),\mathbb{R})}, \|y_1 - y_2\|_{L^1((0,+\infty),\mathbb{R})}\},$$

where $y_1, y_2 \in Y$. Using the Lebesgue dominated convergence theorem [Rudin (1987), Chap. 1, Th. 1.34], we deduce that the metric space $\mathcal{N} = (Y, d)$ is complete.

Consider the integral operator $T : \mathcal{N} \to C(\mathbb{R}_+, \mathbb{R})$ given by the right-hand member of (4.57). The following estimates

$$|T(y)(t)| \leq |C(t)|(1 + \|y\|_{L^1((0,+\infty),\mathbb{R})}) + C^*(t) \int_t^{+\infty} |y(s)|ds$$

$$\leq C^*(t)(1 + 2\|y\|_{L^1((0,+\infty),\mathbb{R})}) \leq C^*(t) \cdot \left(1 + 2\gamma \int_0^{+\infty} C^*(s)ds\right)$$

$$\leq \gamma \cdot C^*(t), \quad t \geq 0,$$

show that $T : \mathcal{N} \to \mathcal{N}$ is well-defined.

Now, we have

$$|T(y_1)(t) - T(y_2)(t)| \leq C^*(t)\|y_1 - y_2\|_{L^1((0,+\infty),\mathbb{R})}$$

$$+ \int_0^{+\infty} |C(s)|ds \cdot \|y_1 - y_2\|_{L^\infty((0,+\infty),\mathbb{R})}$$

$$\leq (\|C\|_{L^\infty((0,+\infty),\mathbb{R})} + \|C\|_{L^1((0,+\infty),\mathbb{R})}) \cdot d(y_1, y_2),$$

by noticing that $C^*(0) = \|C\|_{L^\infty((0,+\infty),\mathbb{R})}$, and also

$$\int_t^{+\infty} |T(y_1)(s) - T(y_2)(s)|ds \leq \int_t^{+\infty} (|C(s)| \cdot \|y_1 - y_2\|_{L^1})ds$$

$$+ \int_t^{+\infty} C^*(s) \int_s^{+\infty} |y_1(\tau) - y_2(\tau)|d\tau ds$$

$$\leq 2 \int_0^{+\infty} C^*(s)ds \cdot d(y_1, y_2), \quad t \geq 0,$$

which lead to

$$d(T(y_1), T(y_2)) \leq \max\big\{\|C\|_{L^\infty((0,+\infty),\mathbb{R})} + \|C\|_{L^1((0,+\infty),\mathbb{R})},$$

$$2\|C^*\|_{L^1((0,+\infty),\mathbb{R})}\big\} \cdot d(y_1, y_2)$$

$$\leq k_1 d(y_1, y_2),$$

where $y_1, y_2 \in \mathcal{N}$.

Notice that we haven't employed (4.52). To do so, let us use different estimates, namely

$$|\mathcal{T}(y_1)(t) - \mathcal{T}(y_2)(t)| \le |C(t)| \cdot \|y_1 - y_2\|_{L^1((0,+\infty),\mathbb{R})}$$
$$+ \left[\int_t^{+\infty} |C(s)|^2 ds\right]^{\frac{1}{2}} \cdot \left[\int_t^{+\infty} |y_1(s) - y_2(s)|^2 ds\right]^{\frac{1}{2}}$$

and

$$\int_t^{+\infty} |y_1(s) - y_2(s)|^2 ds \le \sup_{\tau \ge 0} |y_1(\tau) - y_2(\tau)| \cdot \int_t^{+\infty} |y_1(s) - y_2(s)| ds$$
$$\le [d(y_1, y_2)]^2, \quad t \ge 0.$$

They imply

$$|\mathcal{T}(y_1)(t) - \mathcal{T}(y_2)(t)| \le \left[|C(t)| + \left(\int_t^{+\infty} |C(s)|^2 ds\right)^{\frac{1}{2}}\right] \cdot d(y_1, y_2)$$
$$\le (\|C\|_{L^\infty((0,+\infty),\mathbb{R})} + \|C\|_{L^2((0,+\infty),\mathbb{R})}) d(y_1, y_2)$$

and

$$\int_t^{+\infty} |\mathcal{T}(y_1)(s) - \mathcal{T}(y_2)(s)| ds \le \Big[\|C\|_{L^1((0,+\infty),\mathbb{R})}$$
$$+ \int_0^{+\infty} \left(\int_t^{+\infty} |C(s)|^2 ds\right)^{\frac{1}{2}} dt\Big]$$
$$\times d(y_1, y_2),$$

thus leading to

$$d(\mathcal{T}(y_1), \mathcal{T}(y_2)) \le \max\big\{\|C\|_{L^\infty((0,+\infty),\mathbb{R})} + \|C\|_{L^2((0,+\infty),\mathbb{R})},$$
$$\|C\|_{L^1((0,+\infty),\mathbb{R})} + \|E\|_{L^1((0,+\infty),\mathbb{R})}\big\} \cdot d(y_1, y_2)$$
$$\le k_2 d(y_1, y_2),$$

where $y_1, y_2 \in \mathcal{N}$.

The operator $\mathcal{T} : \mathcal{N} \to \mathcal{N}$ being a contraction, its fixed point y_0 is the solution of (4.57) we are looking for.

The proof is complete. □

Suppose now that the fractional differential equation (4.19) has a solution $x \in C^1([0, +\infty), \mathbb{R})$, with the formula $x(t) = 1 - \int_t^{+\infty} y(s) ds$ for all $t \ge 0$, which satisfies the restrictions (4.49).

Via (1.58), the function x must verify the identity

$$x'(t) = y(t) = -\frac{1}{\Gamma(\alpha)} \int_0^t \frac{a(s)x(s)}{(t-s)^{1-\alpha}} ds = -\frac{1}{\Gamma(\alpha)} \int_0^t \frac{a(s)}{(t-s)^{1-\alpha}} ds$$

$$+ \frac{1}{\Gamma(\alpha)} \int_0^t \frac{a(s)}{(t-s)^{1-\alpha}} \left(\int_s^t + \int_t^{+\infty} \right) y(\tau) d\tau ds$$

$$= -C(t) \left(1 - \int_t^{+\infty} y(s) ds \right) \tag{4.58}$$

$$+ \int_0^t C(t, \tau) y(\tau) d\tau, \quad t \geq 0. \tag{4.59}$$

We have rescaled C in (4.58) as $C(t) = \frac{1}{\Gamma(\alpha)} \int_0^t \frac{a(s)}{(t-s)^{1-\alpha}} ds$, $t \geq 0$. In (4.59), $C(t, \tau) = \frac{1}{\Gamma(\alpha)} \int_0^\tau \frac{a(s)}{(t-s)^{1-\alpha}} ds$, where $t \geq \tau \geq 0$.

The (still unsettled) issue of "$x' \in L^1$" is recast as the statement: *can one solve the integral equation*

$$y(t) = -C(t) \left(1 - \int_t^{+\infty} y(s) ds \right) + \int_0^t C(t, \tau) y(\tau) d\tau, \quad t \geq 0, \tag{4.60}$$

in $L^1 \cap L^\infty$?

We have only established [Baleanu et al. (2011b)] that, if y is a solution of (4.57) and we assume that $0 = y(0)$ — which will lead to $\int_0^{+\infty} (Cy)(s) ds = 0$ by taking $t = 0$ in (4.57) —, and we replace $C(t, \tau)$ in (4.59) with $C(\tau, \tau) = C(\tau)$, then the integral expression from (4.60) would become

$$y(t) = -C(t) \left(1 - \int_t^{+\infty} y(s) ds \right) - \int_t^{+\infty} (Cy)(s) ds, \quad t \geq 0,$$

which is exactly (4.57).

4.5 L^p–Solutions of the Equation (4.3)

A fundamental problem in the spectral theory of ordinary differential equations, namely the *limit–point/limit–circle classification*, see [Weyl (1910)], [Kauffman et al. (1977), Chapter 3], led Wintner [Wintner (1950)] to establish in 1950 that the linear differential equation

$$x''(t) + a(t)x(t) = 0, \quad t \geq t_0 \geq 0, \tag{4.61}$$

where the coefficient $a : [t_0, +\infty) \to \mathbb{R}$ is continuous and such that

$$\int_{t_0}^{+\infty} t^3 |a(t)|^2 dt < +\infty, \tag{4.62}$$

has no non-trivial classical (C^2) solution in L^2.

The result is extremely sharp, see [Bartušek et al. (2004), pp. 59–60] or [Grammatikopoulos and Kulenović (1981), p. 135], and has been followed by many studies about the non-existence of L^2 or L^p–solutions, where $p > 1$, for linear and nonlinear ordinary differential equations of various orders. We mention only a few such references where methods compatible with those in the following have been used, namely [Burlak (1965); Graef and Spikes (1987); Grammatikopoulos and Kulenović (1981); Patula and Waltman (1974)]. The references [Mustafa and Rogovchenko (2003a, 2004b)] and the authoritative monograph [Bartušek et al. (2004)] contain surveys of the abundent literature on this topic.

Here, we are interested in the existence of any fractional variants of the above theorem due to Wintner.

Consider the initial value problem for the fractional differential equation displayed below

$$\begin{cases} {}^2_0\mathcal{O}^{1+\alpha}_t(x)(t) + a(t)x(t) = 0,\, t > 0, \\ \lim_{t \searrow 0}[t^{1-\alpha}x(t)] = 0, \end{cases} \tag{4.63}$$

where the functional coefficient a is continuous in $(0, +\infty)$. Consider also that $\alpha \in \left(\frac{1}{p}, \frac{1}{q}\right)$, where $p > q > 1$ and $\frac{1}{p} + \frac{1}{q} = 1$.

To understand the presence of the initial datum in (4.63), let us emphasize some of the features of the proofs from the references. In the "classical" case (4.61), the restriction

$$\int_{t_0}^{+\infty} |a(t)|^2 dt < +\infty \tag{4.64}$$

will make any (presumable) L^2–solution $x(t)$ be confined to the class $\lim_{t \to +\infty}$ $x'(t) = 0$, see [Wintner (1950), p. 249]. Also, a further restriction

$$\int_{t_0}^{+\infty} t^2 |a(t)|^2 dt < +\infty \tag{4.65}$$

yields

$$\lim_{t \to +\infty} x(t) = \lim_{t \to +\infty} x'(t) = 0. \tag{4.66}$$

Simple examples, see [Mustafa and Rogovchenko (2004b), p. 557], show that the restriction (4.65) alone cannot remove from the solution set of (4.61) the non-trivial L^2–solutions. Up to this day, all the improvements of Wintner's effective hypothesis (4.62) proved to be extremely complicated, see [Grammatikopoulos and Kulenović (1981); Mustafa and Rogovchenko (2004b)].

Various techniques for reconstructing the class (4.66) from the form of the differential equation by means of different hypotheses like (4.64), (4.65) and, in particular, a review of the Barbălat lemma used in control theory can be read in [Mustafa and Rogovchenko (2004a)]. Consequently, it has become clear to us that some *fractional variant* of (4.66) must be added to the fractional differential equation under investigation.

We opted here for inserting the fractional reformulation of the first half of (4.66) in the set of hypotheses. This choice is in line with the classical case, see [Mustafa and Rogovchenko (2004b), Theorem 3]. The second half of (4.66) is retrieved via the next result.

Lemma 4.3. *Suppose that*

$$\int_1^{+\infty} |a(t)|^q dt < +\infty. \tag{4.67}$$

Then, if $x \in \mathcal{RL}^{1-\alpha}$ *is a (mild) solution of problem (4.63) and also* $x \in L^p((1, +\infty), \mathbb{R})$, *we have*

$$\lim_{t \to +\infty} ({}_0D_t^\alpha x)(t) = 0. \tag{4.68}$$

Proof. Given the numbers $t_2 \geq t_1 \geq 1$, we deduce via the Hölder inequality that

$$|({}_0D_t^\alpha x)(t_2) - ({}_0D_t^\alpha x)(t_1)| \leq \int_{t_1}^{t_2} |a(t)x(t)| dt$$

$$\leq \left(\int_{t_1}^{t_2} |a(t)|^q dt \right)^{\frac{1}{q}} \left(\int_{t_1}^{+\infty} |x(t)|^p dt \right)^{\frac{1}{p}}$$

$$\leq \left(\int_1^{+\infty} |a(t)|^q dt \right)^{\frac{1}{q}} \left(\int_{t_1}^{+\infty} |x(t)|^p dt \right)^{\frac{1}{p}}.$$

By means of (4.67), we obtain that the latter quantity tends to zero when $t_1 \to +\infty$. This implies that $\lim_{t \to +\infty} ({}_0D_t^\alpha x)(t) = l \in \mathbb{R}$.

Suppose for the sake of contradiction that $l \neq 0$. Thus, using the L'Hôpital rule and the definition (1.5) of the Riemann-Liouville derivative, we get that

$$\int_0^t \frac{|x(s)|}{(t-s)^\alpha} ds \geq \left| \int_0^t \frac{x(s)}{(t-s)^\alpha} ds \right| \geq \Gamma(1-\alpha) \frac{|l|}{2} \cdot t, \quad t \geq T,$$

for some $T > 1$ large enough.

Also,

$$\int_0^t \frac{|x(s)|}{(t-s)^\alpha} ds \leq B(\alpha, 1-\alpha) \cdot \sup_{s \in (0,1]} \left[s^{1-\alpha} |x(s)| \right] + \int_1^t \frac{|x(s)|}{(t-s)^\alpha} ds,$$

$$= O(1) + \int_1^t \frac{|x(s)|}{(t-s)^\alpha} ds, \quad t \geq 1.$$

Another application of the Hölder inequality leads to

$$
\Gamma(1-\alpha)\frac{|l|}{2}\cdot t \leq \left(\int_1^t \frac{ds}{(t-s)^{\alpha q}}\right)^{\frac{1}{q}}\left(\int_1^{+\infty}|x(s)|^p ds\right)^{\frac{1}{p}}+O(1)
$$

$$
\leq \left(\int_0^t \frac{ds}{(t-s)^{\alpha q}}\right)^{\frac{1}{q}}\left(\int_1^{+\infty}|x(s)|^p ds\right)^{\frac{1}{p}}+O(1)
$$

$$
= \frac{t^{\frac{1}{q}-\alpha}}{(1-\alpha q)^{\frac{1}{q}}}\cdot \|x\|_{L^p((1,+\infty),\mathbb{R})}+O(1), \quad t \geq T.
$$

We have arrived at a contradiction, namely

$$
\|x\|_{L^p((1,+\infty),\mathbb{R})} \geq \lambda \cdot t^{\frac{1}{p}+\alpha} \longrightarrow +\infty \quad \text{when } t \to +\infty,
$$

where $\lambda = \frac{1}{4}(1-\alpha q)^{\frac{1}{q}}\cdot \Gamma(1-\alpha)$.

The proof is complete. $\square$

Assuming that x is an L^p–solution of (4.63) from Lemma 4.3, we deduce that it verifies the integro-differential equation (of fractional order)

$$
(_0D_t^\alpha x)(t) = \int_t^{+\infty} a(s)x(s)ds, \quad t > 0,
$$

which can be recast as a singular integral equation

$$
x(t) = \frac{1}{\Gamma(1-\beta)}\int_0^t \frac{1}{(t-s)^\beta}\int_s^{+\infty} a(\tau)x(\tau)d\tau ds, \quad t > 0, \quad (4.69)
$$

where $\beta = 1 - \alpha$, recall (1.62). Since $\alpha \in \left(\frac{1}{p}, \frac{1}{q}\right)$, we have also $\beta \in \left(\frac{1}{p}, \frac{1}{q}\right)$. According to Theorem 4.1, *the solutions of the linear differential equation under investigation resemble t^α and $t^{-\beta}$ for all the large values of t*. Since, obviously, $t^{-\beta} \in L^p((1,+\infty),\mathbb{R})$, to stay away from such solutions we have imposed that $x_0 = 0$. Thus, the equation (4.69) is the integral reformulation of the boundary value problem

$$
\begin{cases}
{_0^2}\mathcal{O}_t^{1+\alpha}(x)(t) + a(t)x(t), \ t > 0, \\
\lim_{t\searrow 0}\left[t^{1-\alpha}x(t)\right] = x_0, \qquad \lim_{t\to+\infty}(_0D_t^\alpha x)(t) = x_1
\end{cases} \quad (4.70)
$$

in the case of $x_0 = x_1 = 0$.

It is natural now to wonder about the behavior of the (presumable) non-trivial solutions to (4.69). For instance, take $a(t) = \frac{\alpha}{\Gamma(1-\alpha)}\cdot t^{-(1+\alpha)}$ to obtain $x(t) = 1$ everywhere and, for $\varepsilon \in (\alpha, 1)$, $a(t) = \frac{\varepsilon\Gamma(1+\alpha-\varepsilon)}{\Gamma(1-\varepsilon)} \cdot t^{-(1+\alpha)}$ to obtain $x(t) = t^{\alpha-\varepsilon}$, where $t > 0$. These are examples of bounded and respectively eventually bounded solutions to (4.69).

Our first approach to nonexistence is inspired by the classical case (4.61), namely it consists of *reducing* the search for L^p–solutions to one for bounded solutions and then of *giving* a flexible criterion for their nonexistence. See in this respect the paper [Mustafa and Rogovchenko (2004a)] and its references.

Theorem 4.7. *Suppose that the coefficient a verifies the hypothesis*

$$\int_0^1 \frac{|a(\tau)|}{\tau^{1-\alpha}}d\tau + \sup_{t \geq 1}\left[t\int_t^{+\infty}|a(\tau)|^q d\tau\right] < +\infty. \tag{4.71}$$

Then, any non-trivial continuous solution x of the problem (4.63) which lies in $L^p((1, +\infty), \mathbb{R})$ can be expressed as $O\left(t^{-\beta+\frac{1}{p}}\right) = o(1)$ when $t \to +\infty$.

Proof. Notice first that (4.71) implies (4.67) which means that we can use the integral equation (4.69).

Now, let x be a non-trivial solution from $L^p((1, +\infty), \mathbb{R})$ of (4.69). *We claim that* $\int_0^{+\infty}|ax|(\tau)d\tau < +\infty$. To see this, apply the Hölder inequality and notice that $\frac{1}{q} - \beta = \alpha - \frac{1}{p} > 0$:

$$\int_0^{+\infty}|ax|(\tau)d\tau \leq \int_0^1 \frac{|a(\tau)|}{\tau^{1-\alpha}}d\tau \cdot \sup_{s \in (0,1]}\left[s^{1-\alpha}|x(s)|\right]$$

$$+ \left(\int_1^{+\infty}|a(\tau)|^q d\tau\right)^{\frac{1}{q}}\left(\int_1^{+\infty}|x(\tau)|^p d\tau\right)^{\frac{1}{p}} < +\infty.$$

The claim is established.

Further, given $t \geq 1$, we deduce from (4.69) that

$$|x(t)| \leq \frac{1}{\Gamma(1-\beta)}\int_0^t \frac{1}{(t-s)^\beta s^{\frac{1}{q}}} \cdot s^{\frac{1}{q}}\int_s^{+\infty}|ax|(\tau)d\tau ds \leq \frac{1}{\Gamma(1-\beta)}$$

$$\times \int_0^t \frac{ds}{(t-s)^\beta s^{\frac{1}{q}}} \cdot \left\{\int_0^{+\infty}|ax|(\tau)d\tau + \sup_{s \geq 1}\left[s^{\frac{1}{q}}\int_s^{+\infty}|ax|(\tau)d\tau\right]\right\}$$

$$\leq \frac{B\left(\frac{1}{p}, \alpha\right)}{\Gamma(\alpha)} \cdot t^{-\beta+\frac{1}{p}} \cdot I(x),$$

where

$$I(x) = \int_0^{+\infty}|ax|(\tau)d\tau + \sup_{s \geq 1}\left[s^{\frac{1}{q}}\int_s^{+\infty}|ax|(\tau)d\tau\right]$$

$$\leq \int_0^{+\infty} |ax|(\tau)d\tau + \left\{ \sup_{s\geq 1} \left[s \int_s^{+\infty} |a(\tau)|^q d\tau \right] \right\}^{\frac{1}{q}} \cdot \|x\|_{L^p((1,+\infty),\mathbb{R})}$$
$$< +\infty.$$

We have obtained that $x \in L^\infty((1,+\infty),\mathbb{R})$. The proof is complete. $\square$

The next result is about the eventually bounded solutions of problem (4.63).

Theorem 4.8. *Suppose that the following Hille type of restriction*

$$\psi = \Gamma(1-\alpha) \cdot \left\{ \int_0^1 \frac{|a(\tau)|}{\tau^{1-\alpha}} d\tau + \int_1^{+\infty} |a(\tau)|d\tau \right.$$

$$+ \sup_{s\geq 1} \left[s^\alpha \int_s^{+\infty} |a(\tau)|d\tau \right] \right\} < 1 \qquad (4.72)$$

holds true. Then, the problem (4.63) has no non-trivial solutions in $L^\infty((1, +\infty),\mathbb{R})$.

Proof. Notice that $a \in L^1((0,+\infty),\mathbb{R})$. By arguments similar to those in the proof of Lemma 4.3, one can establish that $\lim_{t\to+\infty} ({}_0D_t^\alpha x)(t) = 0$ for any (presumable) non-trivial solution $x \in L^\infty((1,+\infty),\mathbb{R})$. In this way, the problem (4.63) can be recast as (4.69).

Introduce the notation

$$\|x\|_\infty = \max \left\{ \sup_{t\in(0,1]} \left[t^\beta |x(t)| \right] , \|x\|_{L^\infty((1,+\infty),\mathbb{R})} \right\}$$

for some eventually bounded, non-trivial solution x of the integral equation (4.69) which is assumed to exist.

Now, for $t \in (0,1]$, we get that

$$|x(t)| \leq \frac{1}{\Gamma(1-\beta)} \int_0^t \frac{ds}{(t-s)^\beta} \cdot \left[\int_0^1 \frac{|a(\tau)|}{\tau^{1-\alpha}} d\tau + \int_1^{+\infty} |a(\tau)|d\tau \right] \cdot \|x\|_\infty$$

$$\leq \frac{t^{1-\beta}}{\Gamma(2-\beta)} \cdot \left[\int_0^1 \frac{|a(\tau)|}{\tau^{1-\alpha}} d\tau + \int_1^{+\infty} |a(\tau)|d\tau \right] \cdot \|x\|_\infty$$

and respectively

$$\sup_{t\in(0,1]} \left[t^\beta |x(t)| \right] \leq \frac{1}{\Gamma(2-\beta)} \left[\int_0^1 \frac{|a(\tau)|}{\tau^{1-\alpha}} d\tau + \int_1^{+\infty} |a(\tau)|d\tau \right] \cdot \|x\|_\infty.$$

$$(4.73)$$

Further, for $t \geq 1$ and $\chi = \frac{B(\beta,1-\beta)}{\Gamma(1-\beta)} = \Gamma(\beta)$, we get that

$$|x(t)| \leq \frac{1}{\Gamma(1-\beta)} \int_0^t \frac{ds}{(t-s)^\beta s^{1-\beta}} \cdot s^{1-\beta} \int_s^{+\infty} |a(\tau)| \cdot |x(\tau)|d\tau ds$$

$$\leq \chi \left\{ \int_0^{+\infty} |ax|(\tau)d\tau + \sup_{s \geq 1} \left[s^{1-\beta} \int_s^{+\infty} |a(\tau)|d\tau \right] \cdot \|x\|_{L^\infty((1,+\infty),\mathbb{R})} \right\}$$

$$\leq \Gamma(\beta) \left\{ \int_0^1 \frac{|a(\tau)|}{\tau^{1-\alpha}} d\tau + \int_1^{+\infty} |a(\tau)|d\tau + \sup_{s \geq 1} \left[s^{1-\beta} \int_s^{+\infty} |a(\tau)|d\tau \right] \right\}$$

$$\times \|x\|_\infty. \tag{4.74}$$

By combining (4.73), (4.74), we get that

$$\|x\|_\infty \leq \xi \cdot \|x\|_\infty,$$

where ξ is the biggest of the two quantities

$$\frac{1}{\Gamma(2-\beta)} \left[\int_0^1 \frac{|a(\tau)|}{\tau^{1-\alpha}} d\tau + \int_1^{+\infty} |a(\tau)|d\tau \right]$$

and

$$\Gamma(\beta) \left\{ \int_0^1 \frac{|a(\tau)|}{\tau^{1-\alpha}} d\tau + \int_1^{+\infty} |a(\tau)|d\tau + \sup_{s \geq 1} \left[s^{1-\beta} \int_s^{+\infty} |a(\tau)|d\tau \right] \right\}.$$

Via (1.4), we deduce that $\xi = \psi$.

Since $\psi < 1$ and x is non-trivial, we have reached a contradiction, namely

$$0 < \|x\|_\infty \leq \psi \cdot \|x\|_\infty.$$

The proof is complete. $\square$

The issue of bounded solutions to (4.69) can be settled similarly.

Proposition 4.4. *Suppose that the restriction*

$$\vartheta = \Gamma(1-\alpha) \cdot \sup_{s>0} \left[s^\alpha \int_s^{+\infty} |a(\tau)|d\tau \right] < 1 \tag{4.75}$$

holds true. Then, the problem (4.63) has no non-trivial bounded solutions.

The example from page 112 regarding the solution $x(t) = 1$ (everywhere) of (4.69) emphasizes the sharpness of the estimate (4.75).

To conclude this approach, we state the following theorem.

Theorem 4.9. ([Baleanu et al. (2011c), Theorem 3]) *Assume that (4.71), (4.72) are valid. Then, there is no non-trivial solution in $L^p((1,+\infty),\mathbb{R})$ of the problem (4.63).*

Our second approach to nonexistence is based on *comparing* the (presumable) L^p–solutions of (3.40) with the eventually large functions t^α when $t \to +\infty$.

Theorem 4.10. *Suppose that (4.7) holds true for $T = 1$, that is*

$$\eta = \frac{1}{\Gamma(1+\alpha)} \left[\int_0^1 \frac{|a(\tau)|}{\tau^{1-\alpha}} d\tau + \int_1^{+\infty} \tau^\alpha |a(\tau)| d\tau \right] < 1. \qquad (4.76)$$

Then, the problem (4.63) has no non-trivial solution $x \in L^p((1,+\infty),\mathbb{R})$ with the long-time behavior expressed as

$$x(t) = O(t^\alpha) \quad when \ t \to +\infty.$$

Proof. Assuming that such a solution x exists, introduce the notation

$$\|x\|^\infty = \max \left\{ \sup_{s \in (0,1]} \left[s^\beta |x(s)| \right], \sup_{s \geq 1} \frac{|x(s)|}{s^\alpha} \right\}.$$

Further, given the numbers $t_2 \geq t_1 \geq 1$, we have that

$$\begin{aligned}
|(_0D_t^\alpha x)(t_2) - (_0D_t^\alpha x)(t_1)| &\leq \int_{t_1}^{t_2} \tau^\alpha |a(\tau)| \cdot \frac{|x(\tau)|}{\tau^\alpha} d\tau \\
&\leq \int_{t_1}^{+\infty} \tau^\alpha |a(\tau)| d\tau \cdot \|x\|^\infty \\
&= o(1) \quad when \ t_1 \to +\infty.
\end{aligned}$$

Following verbatim the computations from Lemma 4.3, we obtain (4.68) which means that x verifies the integral equation (4.69).

Similarly to the proof of Theorem 4.8, we get also that

$$|x(t)| \leq \frac{t^{1-\beta}}{\Gamma(2-\beta)} \left[\int_0^1 \frac{|a(\tau)|}{\tau^{1-\alpha}} d\tau + \int_1^{+\infty} \tau^\alpha |a(\tau)| d\tau \right] \cdot \|x\|^\infty$$

for every $t > 0$, and respectively $\|x\|^\infty \leq \eta \cdot \|x\|^\infty$.
The proof is complete. $\square$

Theorem 4.11. ([Baleanu et al. (2011c), Theorem 5]) *Let the hypotheses (4.67) and (4.76) be satisfied. Then, the problem (4.63) has no non-trivial solution in $L^p((1,+\infty),\mathbb{R})$.*

Proof. Suppose for the sake of contradiction that there is such a nontrivial solution x of (4.63) which lies in $L^p((1,+\infty),\mathbb{R})$. By taking into account Lemma 4.3, we get that this solution verifies the integral equation (4.69).

Via the Hölder inequality, we establish that

$$|x(t)|$$
$$\leq \frac{1}{\Gamma(1-\beta)} \int_0^t \frac{ds}{(t-s)^\beta} \cdot \int_0^{+\infty} |ax|(\tau)d\tau \leq \frac{t^{1-\beta}}{\Gamma(2-\beta)}$$
$$\times \left\{ \left[\int_0^1 \frac{|a(\tau)|}{\tau^{1-\alpha}} d\tau \cdot \sup_{s \in (0,1]} \left[s^{1-\alpha}|x(s)| \right] + \|a\|_{L^q((1,+\infty),\mathbb{R})} \cdot \|x\|_{L^p((1,+\infty),\mathbb{R})} \right\}$$
$$= O(t^\alpha) \quad \text{when } t \to +\infty.$$

The conclusion follows from Theorem 4.10. $\square$

To conclude this chapter, let us present two results concerning the boundary value problem (4.70).

Proposition 4.5. *Assume that $\int_0^1 \frac{|a(\tau)|}{\tau^{1-\alpha}} d\tau < +\infty$ and*

$$\int_0^{+\infty} \tau^\alpha |a(\tau)| d\tau < \frac{\Gamma(1+\alpha)}{2}.$$

Then, the problem (4.70) with $x_0 \neq 0$ and $x_1 = 0$ has a non-trivial solution in $L^p((1, +\infty), \mathbb{R})$.

Proof. Let X be the set of all the functions $x \in C((0, +\infty), \mathbb{R})$ such that $\lim_{t \searrow 0} \left[t^{1-\alpha} x(t) \right] = l_x \in \mathbb{R}$ and

$$\left| x(t) - x_0 \cdot t^{\alpha-1} \right| \leq \varepsilon |x_0| \cdot t^{\alpha-1}, \quad t > 0,$$

for some $\varepsilon \in (0, 1)$ such that

$$\frac{1+\varepsilon}{\varepsilon} \int_0^{+\infty} s^\alpha |a(s)| ds < \Gamma(1+\alpha).$$

Obviously, $l_x \in [x_0 - \varepsilon|x_0|, x_0 + \varepsilon|x_0|]$. Consider also the distance

$$d(x, y) = \sup_{t > 0} \frac{|x(t) - y(t)|}{t^{\alpha-1}}, \quad x, y \in X.$$

It is easy to see that $\mathcal{X} = (X, d)$ is a complete metric space.

Notice also that

$$\int_0^{+\infty} |ax|(s)ds \leq \int_0^{+\infty} \frac{|a(s)|}{s^{1-\alpha}} ds \cdot \sup_{s > 0} \left[s^{1-\alpha}|x(s)| \right]$$
$$\leq \left[\int_0^1 \frac{|a(s)|}{s^{1-\alpha}} ds + \int_1^{+\infty} s^\alpha |a(s)| ds \right] \cdot \sup_{s > 0} \left[s^{1-\alpha}|x(s)| \right]$$
$$\leq \left[\int_0^1 \frac{|a(s)|}{s^{1-\alpha}} ds + \int_1^{+\infty} s^\alpha |a(s)| ds \right] \cdot (1 + \varepsilon)|x_0|$$
$$< +\infty.$$

Introduce the integral operator $\mathcal{T} : X \to C((0, +\infty), \mathbb{R})$ with the formula

$$\mathcal{T}(x)(t)$$

$$= x_0 t^{\alpha-1} + \frac{1}{\Gamma(\alpha)} \int_0^t \frac{1}{(t-s)^{1-\alpha}} \int_s^{+\infty} (ax)(\tau) d\tau ds$$

$$= x_0 t^{\alpha-1} + \frac{1}{\Gamma(1+\alpha)} \left\{ \int_0^t [t^\alpha - (t-s)^\alpha](ax)(s) ds + t^\alpha \int_t^{+\infty} (ax)(s) ds \right\},$$

where $x \in X$ and $t > 0$. See (4.10) in this respect.

Now, we get that

$$\left| \mathcal{T}(x)(t) - x_0 t^{\alpha-1} \right|$$

$$\leq \frac{(1+\varepsilon)|x_0|}{\Gamma(1+\alpha)} \left[\int_0^t \frac{t^\alpha - (t-s)^\alpha}{s^{1-\alpha}} |a(s)| ds + t^\alpha \int_t^{+\infty} \frac{|a(s)|}{s^{1-\alpha}} ds \right] \quad (4.77)$$

$$\leq \frac{(1+\varepsilon)|x_0|}{\Gamma(1+\alpha)} t^\alpha \cdot \left[\int_0^t \frac{\frac{s}{t}}{s^{1-\alpha}} |a(s)| ds + \frac{1}{t} \int_t^{+\infty} s^\alpha |a(s)| ds \right]$$

$$\leq \frac{(1+\varepsilon)|x_0|}{\Gamma(1+\alpha)} \int_0^{+\infty} s^\alpha |a(s)| ds \cdot t^{\alpha-1} < \varepsilon |x_0| t^{\alpha-1}, \quad t > 0. \quad (4.78)$$

Taking into account (4.77), we deduce as well that

$$\left| t^{1-\alpha} \mathcal{T}(x)(t) - x_0 \right| \leq t^{1-\alpha} \cdot \frac{(1+\varepsilon)|x_0|}{\Gamma(1+\alpha)} \left[t^\alpha \int_0^t \frac{|a(s)|}{s^{1-\alpha}} ds \right.$$

$$\left. + t^\alpha \int_0^{+\infty} \frac{|a(s)|}{s^{1-\alpha}} ds \right]$$

$$\leq t \cdot \frac{(1+\varepsilon)|x_0|}{\Gamma(1+\alpha)} \cdot 2 \int_0^{+\infty} \frac{|a(s)|}{s^{1-\alpha}} ds$$

$$= o(1) \quad \text{when } t \searrow 0, \quad x \in X.$$

So, the operator $\mathcal{T}$ is well-defined: $\mathcal{T}(X) \subseteq X$.

Further, for any $x, y \in X$, we get that

$$|\mathcal{T}(x)(t) - \mathcal{T}(y)(t)| \leq \frac{1}{\Gamma(1+\alpha)} \int_0^{+\infty} s^\alpha |a(s)| ds \cdot t^{\alpha-1} \cdot d(x,y)$$

and respectively that

$$d(\mathcal{T}(x), \mathcal{T}(y)) \leq \frac{1}{2} \cdot d(x,y).$$

The integral operator $\mathcal{T}$ being a contraction of coefficient $\frac{1}{2}$ of the metric space X, the Banach contraction principle yields the existence of a fixed point of $\mathcal{T}$ in X, denoted by x_∞. The conclusion follows from the double inequality

$$\frac{x_0 - \varepsilon|x_0|}{t^\beta} \leq x_\infty(t) \leq \frac{x_0 + \varepsilon|x_0|}{t^\beta}, \quad t \geq 1.$$

The proof is complete. $\square$

Proposition 4.6. *Assume that (4.7) holds true and $\int_T^{+\infty} \tau^{2\alpha}|a(\tau)|d\tau < +\infty$. Then, for any solution x of problem (4.70), one has*

$$\frac{1}{t}\int_T^t \left[t^\alpha - (t-s)^\alpha\right](ax)(s)ds \in L^1((T, +\infty), \mathbb{R}).$$

Proof. If x is a solution of (4.70) then $x(t) = O(t^\alpha)$ as $t \to +\infty$, see [*Baleanu et al. (2010b)*], (4.18).

As before, we have the estimates

$$\frac{1}{t}\left|\int_T^t \left[t^\alpha - (t-s)^\alpha\right](ax)(s)ds\right| \leq \frac{1}{t^{2-\alpha}}\int_T^t s|ax|(s)ds$$

$$\leq \frac{1}{t^{2-\alpha}}\int_T^t s^{1-\alpha} \cdot s^{2\alpha}|a(s)|ds \cdot \sup_{s \geq T}\frac{|x(s)|}{s^\alpha}$$

$$= O\left(\frac{1}{t^{2-\alpha}}\int_T^t s^{1-\alpha} \cdot s^{2\alpha}|a(s)|ds\right) \quad (4.79)$$

when $t \to +\infty$.

Let us make a claim. *If the continuous function $h : [T, +\infty) \to \mathbb{R}$ is in $L^1((T, +\infty), \mathbb{R})$ then $\frac{1}{t^{1+\varepsilon}}\int_T^t s^\varepsilon h(s)ds \in L^1((T, +\infty), \mathbb{R})$ for any $\varepsilon \in (0, 1)$.*

To prove this claim, we use integration by parts

$$\int_T^t \frac{1}{s^{1+\varepsilon}}\int_T^s \tau^\varepsilon h(\tau)d\tau ds = \int_T^t \left(-\int_s^{+\infty}\frac{d\tau}{\tau^{1+\varepsilon}}\right)'\int_T^s \tau^\varepsilon h(\tau)d\tau ds$$

$$= \frac{1}{\varepsilon}\int_T^t h(s)ds - \int_t^{+\infty}\frac{d\tau}{\tau^{1+\varepsilon}} \cdot \int_T^t \tau^\varepsilon h(\tau)d\tau$$

$$= \frac{1}{\varepsilon}\left[\int_T^t h(s)ds - \frac{1}{t^\varepsilon}\int_T^t s^\varepsilon h(\tau)d\tau\right]$$

$$= \frac{1}{\varepsilon}\int_T^t h(s)ds + o(1) \quad \text{when } t \to +\infty,$$

see [Baleanu et al. (2010b)] or the claim from page 84.

The claim is established. The conclusion follows from (4.79) by exploiting the claim for $\varepsilon = 1 - \alpha$ and $h(s) = s^{2\alpha}|a(s)|$. $\square$

To interpret the conclusion of Proposition 4.5, notice first that the problem (4.70) with $x_0 \neq 0$ and $x_1 = 0$ can be regarded as a *perturbation* of the boundary value problem (4.63), (4.68). So, the perturbation has a L^p-solution which resembles $t^{-\beta}$ as $t \to +\infty$ but the size of the solution *attenuates* when $x_0 \to 0$, since the latter boundary value problem has no non-trivial L^p-solutions. This is in line with the classical case: in [Mustafa and Rogovchenko (2003b)] it has been established that a perturbation of (4.61) when the Wintner condition (4.62) holds true will produce L^2-solutions which behave like $t^{-\mu}$ as $t \to +\infty$ for some $\mu > \frac{1}{2}$ but they will vanish as the perturbation is taken smaller and smaller.

Chapter 5

Existence and Uniqueness of Solution for Some Delay Differential Equations with Caputo Derivatives

The retarded integral inequalities of Henry-Gronwall type were applied to certain properties of solutions to fractional differential equations with delay [Ye and Gao (2011)].

The nonlocal character of the fractional derivatives, e.g., (1.5), (1.59), makes them very appealing for the mathematical modeling of various physical or social phenomena with retarded feed-back. Traditionally, these phenomena are described using *delay differential equations*. Numerous details about such equations can be read in the monographs [Driver (1977); Hale and Verduyn Lunel (1993)].

On the other hand, the *lateral* — that is, either left or right — *Caputo derivatives* have been employed recently for fractional variational principles, and the fractional control theory (see for example [Agarwal (2002); Klimek (2001); Eqab et al. (2007); Baleanu and Avkar (2004); Baleanu and Agrawal (2006); Baleanu (2009); Baleanu et al. (2012a); Baleanu and Trujillo (2010); Klimek (2009)] and the references therein). Combining the retarded, neutral or advanced differential equations[1] with the fractional derivatives is therefore, see [Deng et al. (2007); Abdeljawad et al. (2008b)], worthy of investigation from both the theoretical and the applied viewpoints.

Given $J = [a, b]$, where $a < b$, let us introduce the *Riemann-Liouville integrals* [Kilbas et al. (2006)] of *order* α on J, namely

$$I_{a+}^{\alpha}(f)(t) = \frac{1}{\Gamma(\alpha)} \int_a^t \frac{f(s)}{(t-s)^{1-\alpha}} ds, \quad t \in J,$$

and respectively

$$I_{b-}^{\alpha}(f)(t) = \frac{1}{\Gamma(\alpha)} \int_t^b \frac{f(s)}{(s-t)^{1-\alpha}} ds.$$

---

[1] All of these equations are generically referred to as differential equations with *deviating arguments*, cf. [El'sgol'tz (1966); Hale (1971)].

Recalling (1.52), notice that $I_{a+}^{\alpha} = \left(\frac{d}{dt}\right)^{-1} \circ {}_{a}D_t^{1-\alpha}$.

Consider now the linear isometry $Q : C(J, \mathbb{R}) \to C(J, \mathbb{R})$ with the formula $Q(f)(t) = f(a + b - t)$, where $t \in J$. It is obvious that[2], by means of the affine change of variables $v = a + b - s$, we get

$$
\begin{aligned}
\left(Q \circ I_{a+}^{\alpha}\right)(f)(t) &= \frac{1}{\Gamma(\alpha)} \int_a^{a+b-t} \frac{f(s)}{(a+b-t-s)^{1-\alpha}} ds \\
&= \frac{1}{\Gamma(\alpha)} \int_t^b \frac{f(a+b-v)}{(v-t)^{1-\alpha}} dv \\
&= \left(I_{b-}^{\alpha} \circ Q\right)(f)(t).
\end{aligned}
\tag{5.1}
$$

Set the integer $m \geq 1$. Employing the customary notations, remember that the set $C(J) = C(J, \mathbb{R}^m)$, of all the continuous functions $f : J \to \mathbb{R}^m$, endowed with the usual numerical operations (on each of the components $\text{proj}_i \circ f : J \to \mathbb{R}$) can be given a Banach space structure by means of the norm [Kilbas et al. (2006); Driver (1977)]

$$
\|x\|_{\infty} = \max_{t \in J} \|x(t)\|, \quad x \in C(J),
$$

where $\| \star \|$ denotes any convenient Euclidean norm on $\mathbb{R}^m$. For an integer $r \geq 0$, $C^r(J) = C^r(J, \mathbb{R}^m)$ represents the real linear space of all the functions $f \in C(J)$ which are r–times continuously differentiable. Here, $C^0(J) = C(J)$. A Banach space structure on $C^r(J)$ is provided by the norm

$$
\|x\|_r = \sum_{k=0}^{r} \|x^{(k)}\|_{\infty},
\tag{5.2}
$$

see [Corduneanu (1973); Walter (1998); Agarwal et al. (2007b)].

Furthermore, if D is a closed subset of $\mathbb{R}^m$, then by $C^r(J, D)$ we refer to the complete metric space of all the functions $f \in C^r(J)$ taking values in D, with the metric

$$
d(f, g) = \|f - g\|_r, \quad f, g \in C^r(J, D).
$$

The condition of *closedness* for D is necessary. To see this, take e from the closure of D and, given the separability of $\mathbb{R}^m$, let $(e_r)_{r \geq 1}$ be a sequence of elements of D converging to e as $r \to +\infty$. The sequence $(f_r)_{r \geq 1}$, with $f_r(t) = e_r$ for any $t \in J$, is a Cauchy sequence in the metric space $(C^r(J, D), d)$. For the latter space to be complete, it is mandatory that the

---

[2]Given the linearity of operator Q, we shall use freely the symbol QI_{a+}^{α} when referring to $Q \circ I_{a+}^{\alpha}$.

limit of the sequence, namely the function $f : J \to \mathbb{R}^m$ with $f(t) = e$ everywhere in J, be an element of the set $C^r(J, D)$. Consequently, $e \in D$ and so D has to be closed.

Once the metric d has been introduced on the set $C^r(J, D)$, we can use it freely to create a convenient framework for the Banach contraction principle.

Given $t, \tau \in \mathbb{R}$, with $\tau \geq 0$, take $J = [t - \tau, t]$, and introduce the $\star_t$- and respectively $\star^t$-operators via the mappings $x \mapsto x_t$ and $x \mapsto x^t$, where

$$x_t(\theta) = x(t + \theta), \quad x^t(\theta) = x(t - \theta), \quad \theta \in [-\tau, 0],$$

see [Hale (1971); Abdeljawad et al. (2008a)].

The following lemma comprises several elementary facts.

Lemma 5.1. *(i) If one of the limits* $\lim\limits_{t \nearrow b} (b-t)^\gamma f(t)$ *and* $\lim\limits_{t \searrow a} (t-a)^\gamma f(a+b-t)$
exists for some $f \in C([a, b))$ *and* $\gamma \in [0, 1]$, *then both of them exist and are equal.*
(ii) $x_{b+a-t} = (Qx)^t$ *when* $x \in C([a-\tau, b])$. *In particular, we obtain* $x_a = z^b$,
where $z = Qx$.
(iii) $x^{b+a-t} = (Qx)_t$, $x \in C([a, b+\tau])$. *As before,* $x^b = z_a$.

In a loose manner, we shall refer to the quantity $(t - a)^\gamma f(a + b - t)$ from the lemma as $Q((b - t)^\gamma f)$ in the sequel.

To generalize (1.59), fix both an integer $n \geq 1$ and $\alpha \in (0, 1)$. Then, for $m = 1$,

$$_aC_t^\alpha(f^{(n-1)})(t) = \frac{1}{\Gamma(1 - \alpha)} \int_a^t \frac{f^{(n)}(s)}{(t - s)^\alpha} ds, \quad f \in C^n([a, b]).$$

Recall that the *floor* $\lfloor u \rfloor$ of some $u \in \mathbb{R}$ stands for the largest integer which is not greater than the number u. Similarly, the *ceiling* of u, denoted $\lceil u \rceil$, is the smallest integer which is not less than u. Now, we can reason as follows: given $f \in C([a, b])$, suppose that there exists the integer $k \geq 0$ such that $f^{(k+1)} \in C([a, b])$. Then, if $u \in (k, k + 1)$, the next quantity makes sense

$$\frac{1}{\Gamma(1 - \{u\})} \int_a^t \frac{f^{(k)}(s)}{(t - s)^{\{u\}}} ds, \quad \text{where } \{u\} = u - \lfloor u \rfloor.$$

Evidently, the restrictions on f can be relaxed by asking that $f^{(k+1)} \in \mathcal{RL}^{1-\{u\}}$ — we still rely on the Abel computation —.

Now, for any $\alpha \in (n, n + 1)$, where $n \geq 0$ is some integer, the *Caputo differential operator* of order α reads as

$$_aC_t^\alpha(f)(t) = \frac{1}{\Gamma(1 - \{\alpha\})} \int_a^t \frac{f^{(\lceil \alpha \rceil)}(s)}{(t - s)^{\{\alpha\}}} ds, \quad t \in (a, b],$$

see [Podlubny (1999a), p. 79]. It is clear that $_aC_t^\alpha = I_{a+}^{1-\{\alpha\}} \circ \left(\frac{d}{dt}\right)^{\lceil\alpha\rceil}$. To this Caputo operator we shall refer from now on as the *left* operator.

In a symmetric way, the *right* Caputo operator is displayed as

$$_tC_b^\alpha(f)(t) = \frac{(-1)^{\lceil\alpha\rceil}}{\Gamma(1-\{\alpha\})} \int_t^b \frac{f^{(\lceil\alpha\rceil)}(s)}{(s-t)^{\{\alpha\}}} ds, \quad t \in [a,b),$$

where $Q\left((b-t)^{1-\{\alpha\}}f^{(\lceil\alpha\rceil)}\right) \in \mathcal{RL}^0$.

The next result is based on (5.1).

Lemma 5.2. ([Abdeljawad et al. (2008a), Lemma 1]) *We have* $Q\ _aC_t^\alpha = {}_tC_b^\alpha\ Q$, $Q\ _tC_b^\alpha = {}_aC_t^\alpha\ Q$, *and* $Q^2 = \mathrm{id}$.

Let us present now the *integral counterparts* of the lateral Caputo differential operators. To this end, we start from (1.60), that is

$$f^{(\lceil\alpha\rceil-1)}(x) = f^{(\lceil\alpha\rceil-1)}(a) + \frac{1}{\Gamma(\{\alpha\})} \int_a^x \frac{g(s)}{(x-s)^{1-\{\alpha\}}} ds, \quad x > a, \quad (5.3)$$

with $\alpha > 0$ assumed not to be integer.

An integration leads to

$$f^{(n-2)}(x) = f^{(n-1)}(a) \cdot (x-a) + \frac{1}{\Gamma(\beta)} \int_a^x \int_a^t \frac{g(s)}{(t-s)^{1-\beta}} ds\,dt, \quad x > a,$$

where $n = \lceil\alpha\rceil = 1 + \lfloor\alpha\rfloor$ and $\beta = \{\alpha\}$. Since — recall (1.26) —

$$\int_a^x \int_a^t \frac{g(s)}{(t-s)^{1-\beta}} ds\,dt = \int_a^x \int_s^x \frac{g(s)}{(t-s)^{1-\beta}} dt\,ds = \int_a^x \frac{(x-s)^\beta}{\beta} g(s) ds,$$

we get that

$$f^{(n-2)}(x) = f^{(n-1)}(a)(x-a) + \frac{1}{\Gamma(1+\beta)} \int_a^x (x-s)^\beta g(s) ds, \quad x \in [a,b].$$

Repeating the procedure, we arrive finally at

$$f(x) = \sum_{j=0}^{n-1} f^{(j)}(a) \cdot \frac{(x-a)^j}{j!} + \frac{1}{\Gamma(n-1+\beta)} \int_a^x (x-s)^{n+\beta-2} g(s) ds$$

$$= \sum_{j=0}^{n-1} b_j(x-a)^j + \frac{1}{\Gamma(\alpha)} \int_a^x (x-s)^{\alpha-1} g(s) ds, \quad (5.4)$$

where $x \in [a,b]$ and $b_j = \frac{f^{(j)}(a)}{j!}$ for $0 \le j \le \lfloor\alpha\rfloor$, see [Kilbas et al. (2006)].

As regards the right Caputo operator, we deduce from Lemma 5.2 that the equation $_tC_b^\alpha(f) = g$, which can be recast as $(Q \circ {}_tC_b^\alpha)(f) = Qg$, is equivalent with $_aC_t^\alpha(Qf) = Qg$. Thus, we deduce from (5.3) that

$$(Qf)^{(\lceil\alpha\rceil-1)}(x) = (Qf)^{(\lceil\alpha\rceil-1)}(a) + \frac{1}{\Gamma(\{\alpha\})} \int_a^x \frac{(Qg)(s)}{(x-s)^{1-\{\alpha\}}} ds$$

and respectively $-Q \circ \frac{d^n}{dx^n} = (-1)^n \frac{d^n}{dx^n} \circ Q$ and $y = Qx = a + b - x$ —

$$(-1)^{\lceil \alpha \rceil - 1} f^{(\lceil \alpha \rceil - 1)}(y)$$

$$= (-1)^{\lceil \alpha \rceil - 1} f^{(\lceil \alpha \rceil - 1)}(b) + \frac{1}{\Gamma(\{\alpha\})} \int_a^{Qy} \frac{(Qg)(s)}{(Qy - s)^{1 - \{\alpha\}}} ds$$

$$= (-1)^{\lceil \alpha \rceil - 1} f^{(\lceil \alpha \rceil - 1)}(b) + I_{a+}^{\{\alpha\}}(Qg)(Qy)$$

$$= (-1)^{\lceil \alpha \rceil - 1} f^{(\lceil \alpha \rceil - 1)}(b) + (Q \circ I_{a+}^{\{\alpha\}})(Qg)(y)$$

$$= (-1)^{\lceil \alpha \rceil - 1} f^{(\lceil \alpha \rceil - 1)}(b) + I_{b-}^{\{\alpha\}}(Q^2 g)(y).$$

In other words,

$$f^{(\lceil \alpha \rceil - 1)}(y) = f^{(\lceil \alpha \rceil - 1)}(b) + \frac{(-1)^{\lceil \alpha \rceil - 1}}{\Gamma(\{\alpha\})} \int_y^b \frac{g(s)}{(s - y)^{1 - \{\alpha\}}} ds, \quad y < b,$$

which leads to

$$f(x) = \sum_{j=0}^{n-1} b_j (b - x)^j - \frac{1}{\Gamma(\alpha)} \int_x^b (s - x)^{\alpha - 1} g(s) ds, \qquad (5.5)$$

where $x \in [a, b]$ and $b_j = (-1)^j \frac{f^{(j)}(b)}{j!}$ for $0 \le j \le \lfloor \alpha \rfloor$, see [Abdeljawad et al. (2008a), Theorem 3].

We are now in a perfect position to state the boundary value problem under investigation in this chapter. Let us start by setting $\tau > 0$ and denoting with B the complete metric space $(C([-\tau, 0]), d)$ — a quick reminder: $C(J) = C(J, \mathbb{R}^m)$ for some $m \ge 1$ —.

Given the integer $r \ge 0$, a functional $F : [a, b] \times B \to \mathbb{R}^m$ is said to fulfill the C_r–*condition* if the mapping $t \mapsto F(t, y_t)$ is continuous throughout $[a, b]$ for each $y \in C^r([a - \tau, b])$, whereas F is said to satisfy the C^r *condition* if $t \mapsto F(t, y^t)$ is continuous in $[a, b]$ for any $y \in C^r([a, b + \tau])$, see [Abdeljawad et al. (2008a), p. 2].

Consider now the boundary value problem

$$\begin{cases} {}_tC_b^\alpha \left({}_aC_t^\alpha x \right)(t) + A \cdot {}_tC_b^\alpha(x)(t) = F(t, x_t), \ t \in (a, b], \\ x^{(k)}(b) = c_k, \ \left({}_aC_t^\alpha x \right)^{(k)}(b) = b_k, \qquad\qquad k \in \overline{0, n - 1}, \qquad (5.6) \\ x_a = \phi, \end{cases}$$

where $n = \lceil \alpha \rceil$ and the functional F is assumed to satisfy the C_r–condition with $r \ge 0$, for some A, b_k, $c_k \in \mathbb{R}$ and $\phi \in B$.

To display the problem as a singular integral equation, notice that it can be decomposed as

$$\begin{cases} {}_aC_t^\alpha(x)(t) + Ax(t) = y(t), \\ {}_tC_b^\alpha(y)(t) = F(t, x_t), \end{cases}$$

where $x^{(k)}(b) = c_k$ and $y^{(k)}(b) = b_k + Ac_k$, $0 \le k \le \lceil \alpha \rceil - 1$.

Further, according to (5.5), we get

$$y(t) = \sum_{j=0}^{n-1} B_j(b-t)^j - \frac{1}{\Gamma(\alpha)} \int_t^b (s-t)^{\alpha-1} F(s, x_s) ds, \quad t \in [a,b], \quad (5.7)$$

with $B_j = \frac{(-1)^j}{j!}(b_j + Ac_j)$, see [Abdeljawad et al. (2008a), Lemma 7]. The integral equation is obtained via (5.4), that is

$$x(t) = \sum_{j=0}^{n-1} C_j(t-a)^j + \frac{1}{\Gamma(\alpha)} \int_a^t (t-s)^{\alpha-1}[y(s) - Ax(s)] ds, \quad (5.8)$$

where $C_j = \frac{x^{(j)}(a)}{j!}$.

Since $x_a = \phi$, then $C_j = \frac{\phi^{(j)}(0)}{j!}$ and we must ask that *the function ϕ be $C^{\lceil \alpha \rceil - 1}$ around $\theta = 0$* for the problem to be well-posed. See also the discussion in [Hale (1971), p. 5].

To give a clearer presentation of (5.8), introduce the quantity $I_{\gamma,j}(b,t) = \int_a^t (t-s)^\gamma (b-s)^j ds$ for $t \in [a,b]$, with $\gamma > -1$. Using integration by parts, we deduce that

$$I_{\alpha-1,j}(b,t) = \sum_{p=0}^{k} (-1)^p \frac{j(j-1)\cdots(j-p+\operatorname{sign} p)}{\alpha(\alpha+1)\cdots(\alpha+p)} \cdot (b-a)^{j-p} \quad (5.9)$$

$$+ (-1)^{k+1} \frac{j(j-1)\cdots(j-k)}{\alpha(\alpha+1)\cdots(\alpha+k)} \cdot I_{\alpha+k,j-k-1}(b,t), \quad (5.10)$$

where $0 \le k \le j-1$. In particular, $I_{\alpha+j-1,0}(b,t) = \frac{(t-a)^{\alpha+j}}{\alpha+j}$. Notice that the quantity $I_{\alpha-1,j}(b,t)$ is dominated by $C \cdot (t-a)^{\alpha+j}$ when[3] $t - a \to +\infty$, with $C = \frac{(-1)^j j!}{\alpha(\alpha+1)\cdots(\alpha+j)}$.

Now, inserting (5.7) into (5.8) yields

$x(t)$

$$= \sum_{j=0}^{n-1} C_j(t-a)^j + \frac{1}{\Gamma(\alpha)} \sum_{j=0}^{n-1} B_j I_{\alpha-1,j}(b,t) \quad (5.11)$$

$$- \frac{1}{\Gamma(\alpha)} \int_a^t (t-s)^{\alpha-1} \left[Ax(s) + \frac{1}{\Gamma(\alpha)} \int_s^b (\zeta-s)^{\alpha-1} F d\zeta \right] ds \quad (5.12)$$

$$= X_0 + \sum_{j=0}^{n-1} \left[C_j(t-a)^j + D_j(t-a)^{\alpha+j} \right] - \frac{A}{\Gamma(\alpha)} \int_a^t (t-s)^{\alpha-1} x(s) ds$$

---

[3] As a by-product, $b - a \to +\infty$.

$$-\frac{1}{[\Gamma(\alpha)]^2}\int_a^t\left(\int_s^t+\int_t^b\right)(t-s)^{\alpha-1}(\zeta-s)^{\alpha-1}F(\zeta,x_\zeta)d\zeta ds, \qquad (5.13)$$

where X_0, $D_j \in \mathbb{R}$. We remark that the number X_0 depends on $b-a$ because of (5.9). The terms $(t-a)^{\alpha+j}$ are a consequence of (5.10) for $k = j-1$.

The integral quantity $\mathcal{I}(t)$ from (5.13) can be recast as

$$\begin{aligned}
\mathcal{I}(t) &= \int_a^t\int_s^t (t-s)^{\alpha-1}(\zeta-s)^{\alpha-1}F(\zeta,x_\zeta)d\zeta ds \\
&\quad + \frac{(t-a)^\alpha}{\alpha}\int_t^b(\zeta-s)^{\alpha-1}F(\zeta,x_\zeta)d\zeta \\
&= \int_a^t\int_a^\zeta (t-s)^{\alpha-1}(\zeta-s)^{\alpha-1}F(\zeta,x_\zeta)dsd\zeta \\
&\quad + \int_a^t(t-s)^{\alpha-1}\left[\int_t^b(\zeta-s)^{\alpha-1}F(\zeta,x_\zeta)d\zeta\right]ds \\
&= \int_a^t I_{\alpha-1,\alpha-1}(t,\zeta)F(\zeta,x_\zeta)d\zeta \\
&\quad + \int_a^t(t-s)^{\alpha-1}\left[\int_t^b(\zeta-s)^{\alpha-1}F(\zeta,x_\zeta)d\zeta\right]ds.
\end{aligned}$$

The integral reformulation of the problem (5.6) reads now as

$$\begin{aligned}
x(t) &= X_0 + \sum_{j=0}^{n-1}\left[C_j(t-a)^j + D_j(t-a)^{\alpha+j}\right] \\
&\quad - \frac{A}{\Gamma(\alpha)}\int_a^t(t-s)^{\alpha-1}x(s)ds \\
&\quad - \frac{1}{[\Gamma(\alpha)]^2}\int_a^t I_{\alpha-1,\alpha-1}(t,\zeta)F(\zeta,x_\zeta)d\zeta \\
&\quad - \frac{1}{[\Gamma(\alpha)]^2}\int_a^t(t-s)^{\alpha-1}\left[\int_t^b(\zeta-s)^{\alpha-1}F(\zeta,x_\zeta)d\zeta\right]ds \quad (5.14)
\end{aligned}$$

in $[a,b]$. Also, $x(t) = \phi(t-a)$ for any $t \in [a-\tau,a]$, see [Abdeljawad et al. (2008a), Lemma 4].

The functional $F : [a,b] \times B \to \mathbb{R}^m$ is called $_r$–*Lipschitzian* if it verifies the C_r–condition and there exists $K > 0$ such that — recall (5.2) —

$$\|G(y_1) - G(y_2)\|_\infty \le K\|y_1 - y_2\|_r \quad \text{for all } y_1, y_2 \in C^r([a-\tau,b]),$$

where $G(y)(t) = F(t, y_t)$, $t \in [a, b]$. Symmetrically, the functional F is ^{r}–*Lipschitzian* if it verifies the C^r–condition and there exists $K > 0$ such that

$$\|H(y_1) - H(y_2)\|_\infty \leq K\|y_1 - y_2\|_r \quad \text{for all } y_1,\ y_2 \in C^r([a, b + \tau]),$$

where $H(y)(t) = F(t, y^t)$, $t \in [a, b]$.

Chapter 6

Existence of Positive Solutions for Some Delay Fractional Differential Equations with a Generalized N–Term

The existence of positive solution for fractional differential equations with delay is an interesting issue both from theoretical and applied viewpoints (see for example refs. [Babakhani and Enteghami (2009); Daftardar-Gejji and Babakhani (2004); Babakhani and Daftardar-Gejji (2003, 2006); Sign and Chuanzhi (2006); Daftardar-Gejji (2005); Ye et al. (2007); Belarbi et al. (2006); Zhang (2008)] and references therein). One illustrative example was provided in [Babakhani and Enteghami (2009)], namely the following type of fractional differential equations was analyzed:

$$\mathcal{L}(D)[x(t) - x(0)] = f(t, x_t), \qquad t \in (0, T],$$
$$x(t) = \phi(t) \geq 0, \qquad t \in [-\omega, 0], \tag{6.1}$$

such that

$$\mathcal{L}(D) = D^{a_n} - \sum_{j=1}^{n-1} p_j(t) D^{a_{n-j}}, \qquad 0 < a_1 < \cdots < a_n < 1,$$

$$p_j(t) = \sum_{k=0}^{N_j} a_{jk} t^k, \qquad p_j^{(2m)}(t) \geq 0, \qquad p_j^{(2m+1)}(t) \leq 0 \tag{6.2}$$

for

$$m = 0, 1, \cdots, \left[\frac{N_j}{2}\right], \qquad j = 1, 2, \cdots, n-1.$$

Here, D^{a_j} represents the standard Riemann-Liouville fractional derivative, $T > 0$, $\omega > 0$, $\phi \in C = C([-\omega, 0], \mathbb{R}^+)$ and $f : I \times C \to \mathbb{R}^+$ is a given continuous function, $I = [0, T]$.

Now we investigate the existence of positive solutions for initial non-linear fractional differential equations with finite delay [Babakhani and Baleanu (2011)],

$$\mathcal{L}(D)[x(t) - x(0)] = f(t, x_t), \quad t \in (0, b],$$
$$x(t) = \omega(t), \quad t \in [-\tau, 0], \tag{6.3}$$

equipped with initial conditions

$$x(0) = x_0, [D^{a-n+1} x(t)]_{t=0} = x_{n-1},$$
$$[D^{a-n+j} x(t)]_{t=0} = x_{n-j},$$

such that $\mathcal{L}(D) = D^a - \sum_{j=1}^{n-1} a_j D^{a-j}, a_j > 0$, for all $j, j = 1, 2, \cdots, n - 1$, $n - 1 < a < n, n \in \mathbb{N}, f : (0, b] \times C \to [0, +\infty)$ denotes a given continuous function fulfilling $\lim_{t \to 0} f(t, x_t) = +\infty$ (i.e., f is singular at $t = 0$), where C is the space of continuous functions from $[-\tau, 0]$ to $\mathbb{R}^{>0}$ and $x_t \in C$ defined by $x_t(s) = x(t + s)$ for each $s \in [-\tau, 0)$. In (6.1), we assume that

$$\text{(i) } x_{n-1} \leq 0, \quad \text{(ii) } x_{n-j} \geq \sum_{k=1}^{j-1} a_k x_{k+n-j}, \quad \forall j = 2, 3, \cdots, n-1. \tag{6.4}$$

6.1 The Existence Theorem

Below we prove that the initial value problem (6.1) under the conditions among the initial value (i.e., (6.3) admits a positive solution).

Lemma 6.1. ([Babakhani and Baleanu (2011)]) *Let* $g : (0, b] \to \mathbb{R}$ *be a continuous function and* $\lim_{t \to 0+} g(t) = +\infty$. *If there exists* $\sigma \in (0, 1)$ *such that* $\sigma < \alpha$ *and by letting* $t^\sigma g(t)$ *be a continuous function on* $[0, b]$, *then* $H(t) = I^\alpha t^\sigma g(t)$ *is continuous* $[0, b]$, *where* $n - 1 < \alpha < n$, $n \in \mathbb{N}$.

Proof. *([Babakhani and Baleanu (2011)])* Let us consider $L = \max t^\sigma g(t)$, $t \in [0, b]$. For all $t \in [0, b)$ and for given all $\epsilon > 0$,

$$|H(t + \epsilon) - H(t)| = \frac{1}{\Gamma(\alpha)} \left| \int_0^{t+\epsilon} (t + \epsilon - s)^{\alpha-1} ds - \int_0^t (t - s)^{\alpha-1} ds \right|$$

$$= \frac{1}{\Gamma(\alpha)} \left| \int_0^{t+\epsilon} (t + \epsilon - s)^{\alpha-1} s^{-\sigma} s^\sigma g(s) ds \right.$$

$$\left. - \int_0^t (t - s)^{\alpha-1} s^{-\sigma} s^\sigma g(s) ds \right|$$

$$\leq \frac{L}{\Gamma(\alpha)} \left| \int_0^{t+\epsilon} (t + \epsilon - s)^{\alpha-1} s^{-\sigma} ds \right.$$

$$-\int_0^t (t-s)^{\alpha-1}s^{-\sigma}ds\Bigg|$$

$$= \frac{L}{\Gamma(\alpha)}B(1-\sigma,\alpha)|(t+\epsilon)^{\alpha-\sigma}-t^{\alpha-\sigma}|. \qquad (6.5)$$

Thus, we find out that $\lim_{\epsilon\to 0}|H(t+\epsilon)-H(t)| = 0$. We conclude that a similar result is conclusion for $|H(b-\epsilon)-H(b)|$. $\square$

The aim is to prove that (6.1) is equivalent to an integral equation.

Theorem 6.1. ([Babakhani and Baleanu (2011)]) *Suppose that* $f : (0,b] \times C \to [0,+\infty)$ *is a given continuous function so that* $\lim_{t\to 0}f(t,x_t) = +\infty$ *(i.e., f is singular at $t=0$), where C is the space of continuous functions from $[-\tau,0]$ to $\mathbb{R}^{\geq 0}$ and $x_t \in C$ defined by $x_t(s) = x(t+s)$ for each $s \in [-\tau,0]$. If there exists $\sigma \in (0,1)$ such that $0 < \sigma < \alpha \in (n-1,n)$ and $t^\sigma f(t,x_t)$ is a continuous function on $[0,b]$, then*

$$\left(D^\alpha - \sum_{j=1}^{n-1} a_j D^{\alpha-j}\right)[x(t)-x(0)] = f(t,x_t), \quad t \in (0,b], \qquad (6.6)$$

is equivalent to

$$x(t) = x(0) + \lambda(t) + \mathcal{L}(I)[x(t)-x(0)] + I^\alpha f(t,x_t), \quad t \in [0,b], \qquad (6.7)$$

where

$$\lambda(t) = \sum_{j=1}^{n-1}\lambda_j t^{\alpha-1},$$

$$\lambda_j = \frac{1}{\Gamma(\alpha-j+1)}\left(x_j - \sum_{k=1}^{n-j-1} a_k x_{k|j}\right), \quad j = 1,\ldots,n-2,$$

$$\lambda_{n-1} = \frac{1}{\Gamma(\alpha-n+1)}, \quad \mathcal{L}(I) = \sum_{j=1}^{n-1} a_j I^j x(t). \qquad (6.8)$$

Proof.*([Babakhani and Baleanu (2011)])* From (6.6), we obtain

$$I^\alpha\{D^\alpha[x(t)-x(0)]\} - \sum_{j=1}^{n-1} a_j I^\alpha\left\{D^{\alpha-j}[x(t)-x(0)]\right\}$$

$$= I^\alpha f(t,x_t), \quad t \in (0,b]. \qquad (6.9)$$

Thus, we have

$$I^\alpha\{D^\alpha[x(t)-x(0)]\} = [x(t)-x(0)] - \sum_{j=1}^{n-1}\frac{x_j t^{\alpha-j}}{\Gamma(\alpha-j+1)}, \qquad (6.10)$$

and, for $k = 1, 2, \ldots, n - 1$,

$$I^\alpha \left\{ D^{\alpha-k}[x(t) - x(0)] \right\} = I^k[x(t) - x(0)] - \sum_{j=1}^{n-k-1} \frac{x_{j+k} t^{\alpha-j}}{\Gamma(\alpha - j + 1)}. \quad (6.11)$$

By using the Lemma 6.1, $I^\alpha f(t, x_1) = I^\alpha(t^\sigma t^{-\sigma} f(t, x_t))$ exists and $D^\alpha I^\alpha$ $f(t, x_t) = f(t, x_t)$, as $t^{-\sigma} f(t, x_t)$ is continuous and $I^\alpha f(t, x_t) \in C[0, b]$. Taking into account (6.9), (6.10) together (6.11), (6.6) is equivalent to the integral equation given below

$$x(t) = x(0) + \lambda(t) + \mathcal{L}(I)[x(t) - x(0)] + I^\alpha f(t, x_1), \quad t \in (0, b], \quad (6.12)$$

such that

$$\lambda(t) = \sum_{j=1}^{n-1} \lambda_j t^{\alpha-j},$$

$$\lambda_j = \frac{1}{\Gamma(\alpha - j + 1)} \left(x_j - \sum_{k=1}^{n-j-1} a_k x_{k+j} \right), \quad j = 1, \ldots, n - 2,$$

$$\lambda_{n-1} = \frac{1}{\Gamma(\alpha - n + 1)}, \quad \mathcal{L}(I) = \sum_{j=1}^{n-1} a_j I^j x(t). \quad (6.13)$$

The proof is complete. [Babakhani and Baleanu (2011)]. As a result by using the Theorem 6.1, the other expression of (6.1) becomes

$$x(t) = \begin{cases} x(0) + \lambda(t) + \mathcal{L}(I)[x(t) - x(0)] + I^\alpha f(t, x_t), \, t \in (0, b], \\ \\ x(t) = \omega(t), \qquad\qquad\qquad\qquad\qquad t \in [-\tau, 0]. \end{cases} \quad (6.14)$$

Let us define the function $y : [-\tau, b] \to [0, +\infty)$ as

$$y(t) = \begin{cases} \omega(0), \qquad t \in [0, b], \\ \\ \omega(t) \geq 0, \, t \in [-\tau, 0], \end{cases} \quad (6.15)$$

for each $z \in C([0, b], \mathbb{R})$ with $z(0) = 0$; $\overline{z}$ represents the function defined as

$$\overline{z}(t) = \begin{cases} z(t), \, t \in [0, b], \\ \\ 0, \qquad t \in [-\tau, 0]. \end{cases} \quad (6.16)$$

At this stage we decompose $x(\cdot)$ as $x(t) = \overline{z}(t) + y(t)$, $t \in [-\tau, b]$, which implies that $x_t = \overline{z}_t + y_t$, $t \in [0, b]$. As a result of using the Theorem 6.1, (6.1) is equivalent to the integral equation

$$z(t) = \lambda(t) + \mathcal{L}(I)z(t) + I^\alpha f(t, \overline{z}_t + y_t), \quad t \in [0, b]. \quad (6.17)$$

Set $\Omega = \{z \in C([0,b], \mathbb{R}), z(0) = 0\}$, and for each $z \in \Omega$, let $||z||_b$ be the seminorm in Ω defined as below

$$\| z \|_b = \| z(0) \| + \| z \| = \| z \| = \sup\{|z(t)| : t \in [0,b]\}. \qquad (6.18)$$

Besides Ω represents a Banach space equipped with norm $\| \cdot \|_b$.

Let K be a cone of Ω. $K = \{z \in \Omega : z(t) \geq 0, \ t \in [0,b]\}$ and

$$K^* = \left\{x \in C\left([-\tau, b], \mathbb{R}^{\geq 0}\right) : x(t) = \omega(t) \geq 0, \quad t \in [-\tau, 0]\right\}. \qquad (6.19)$$

The next step is to define the operator $F : K \to K$ by

$$Fz(t) = \lambda(t) + \mathcal{L}(I)z(t) + I^\alpha f(t, \overline{z}_t + y_t), \quad t \in [0,b]. \qquad (6.20)$$

Theorem 6.2. ([Babakhani and Baleanu (2011)]) *Suppose that $f(t, x_t)$, $t \in (0, b], x_t \in C$, is a continuous function and $\lim_{t \to 0^+} f(t, \cdot) = +\infty$. If there exists $\sigma \in (0,1)$ such that $0 < \sigma < \alpha \in (n-1, n)$ and $t^\sigma f(t, x_t)$ is a continuous function on $[0, b]$, then the operator F, defined as (3.12), maps bounded set into bounded sets in K, continuous and completely continuous.*

Proof. *([Babakhani and Baleanu (2011)])* For all $u \in K$, since $Fu(t) = \lambda(t) + \mathcal{L}(I)u(t) + I^\alpha f(t, \overline{u}_t + y_t)$ by Lemma 6.1 and the nonnegativeness of f, we obtain $F : K \to K$.

Using the fact $t^\sigma f(t, x_t)$ is continuous on $[0, b] \times [0, +\infty)$, there exists a positive constant N fulfilling $t^\sigma f(t, x_t) \leq N$. Thus, we obtain

$$I^\alpha f(t, \overline{z}_t + y_t) = I^\alpha (t^{-\sigma} t^\sigma f(t, x_t))$$

$$\leq N I^\alpha t^{-\sigma} = \frac{N}{\Gamma(\alpha)} \int_0^t (t-s)^{\alpha-1} s^{-\sigma} ds \qquad (6.21)$$

$$= \frac{N t^{\alpha-\sigma}}{\Gamma(\alpha)} B(1 - \sigma, \alpha) \leq \frac{N b^{\alpha-\sigma}}{\Gamma(\alpha)} B(1 - \sigma, \alpha).$$

Let $G \subset K$ be bounded, that is, there exists a positive constant L such that $\| Z \|_b \leq L$, for all $z \in C$. From (6.17), for each $z \in G$, we report

$$\| Fz(t) \|$$

$$\leq |\lambda(t)| + |\mathcal{L}(I)z(t)| + |I^\alpha f(t, \overline{z}_t + y_t)|$$

$$\leq \| \lambda(t) \| + \| z \|_b \sum_{j=1}^n \frac{a_j b^j}{\Gamma(j+1)} + \frac{N b^{\alpha-\sigma}}{\Gamma(\alpha)} B(1 - \sigma, \alpha) \qquad (6.22)$$

$$\leq \| \lambda(t) \| + L \sum_{j=1}^n \frac{a_j b^j}{\Gamma(j+1)} + \frac{N b^{\alpha-\sigma}}{\Gamma(\alpha)} B(1 - \sigma, \alpha).$$

As a result, $F(K)$ is bounded.

In the next step we will prove that F is continuous. Let $v_0 \in K$ and $\| v_0 \| = c_0$, if $v \in K$ and $\| v - v_0 \| < l$, then $\| v \| < 1 + c_0$; by the continuity of $t^\sigma f(t, \overline{z}_t + y_t)$, we conclude that $t^\sigma f(t, \overline{z}_t + y_t)$ is uniformly continuous on $[0, b] \times [0, c]$.

Then, for all $\epsilon > 0$ there exists a $\delta > 0$ such that

$$|t^\sigma f(t, \overline{u}_t + y_t) - t^\sigma f(t, \overline{v}_t + y_t)| < \epsilon, \tag{6.23}$$

for all $t \in [0, b]$ and $u, v \in [0, c]$ with $|u - v| < \delta$. It is clear that, if $\| u - v \| < \delta$, then $v_0(t), v(t) \in [0, c]$ and $|v(t) - v_0(t)| < \delta$ for each $t \in [0, b]$.

As a result, we have

$$|t^\sigma f(t, \overline{v}_t + y_t) - t^\sigma f(t, \overline{v_0}_t + y_t)| < \epsilon, \tag{6.24}$$

for all $t \in [0, b]$ and $v \in K$ with $\| v - v_0 \| < \delta$.

For all $t \in [0, b]$, let $u, v \in K$, and $|u(t) - v(t)| < \delta$, we choose $\delta \leq \epsilon (\sum_{j=1}^{n-1} (x_j b^j / \Gamma(j+1)))^{-1}$. From (6.24), we conclude that

$$
\begin{aligned}
|Fu(t) - Fv(t) &\leq \| u - v \|_b \sum_{j=1}^{n-1} \frac{a_j b^j}{\Gamma(j+1)} \\
&\quad + |I^\alpha (f(t, \overline{u}_t + y_t) - f(t, \overline{v}_t + y_t))| \\
&= \| u - v \|_b \sum_{j=1}^{n-1} \frac{a_j b^j}{\Gamma(j+1)} \\
&\quad + |I^\alpha (t^{-\sigma} t^\sigma f(t, \overline{u}_t + y_t) - t^{-\sigma} t^\sigma f(t, \overline{v}_t + y_t))| \\
&\leq \| u - v \|_b \sum_{j=1}^{n-1} \frac{a_j b^j}{\Gamma(j+1)} \\
&\quad + \frac{1}{\Gamma(\alpha)} \max_{0 \leq t \leq b} \int_0^t (t-s)^{\alpha-1} s^{-\sigma} \\
&\quad \times |s^\sigma f(t, \overline{u}_t + y_t) - s^\sigma f(t, \overline{v}_t + y_t)| ds \\
&\leq \| u - v \|_b \sum_{j=1}^{n-1} \frac{a_j b^j}{\Gamma(j+1)} + \frac{\epsilon}{\Gamma(\alpha)} \max_{0 \leq t \leq b} \int_0^1 (t-s)^{\alpha-1} s^{-\sigma} ds \\
&= \| u - v \|_b \sum_{j=1}^{n-1} \frac{a_j b^j}{\Gamma(j+1)} + \frac{\epsilon}{\Gamma(\alpha)} \max_{0 \leq t \leq b} t^{\alpha-\sigma} B(1-\sigma, \alpha) \\
&\leq (1 + b^{\alpha-\sigma}) B(1-\sigma, \alpha) \epsilon.
\end{aligned}
\tag{6.25}
$$

Our goal is to show that the operator F is equicontinuous [Babakhani and Baleanu (2011)]. Let $G \subseteq K$ be bounded, there exists a positive constant

I such that $\parallel u \parallel \leq l$, for all $u \in G$. Let us suppose, $u \in K, t, r \in [0, b]$ and $t < r$. For a given $\epsilon > 0$, there exists $\delta > 0$, so that if $|t - r| < \delta$, then

$$
|Fu(t) - Fu(r)|
$$

$$
\leq |\lambda(t) - \lambda(r)| + |\mathcal{L}(I)u(t) - \mathcal{L}(I)u(r)|
$$

$$
+ |I^{\alpha} f(t, \overline{u}_t + y_t) - I^{\alpha} f(r, \overline{u}_r + y_r)|
$$

$$
\leq \sum_{j=1}^{n-1} \lambda_j |t^{\alpha-j} - r^{\alpha-j}| + \sum_{j=1}^{n-1} \frac{la_j}{\Gamma(j)} \left| \int_0^1 (t-s)^{j-1} - (r-s)^{j-1} ds \right|
$$

$$
+ \sum_{j=1}^{n-1} \frac{la_j}{\Gamma(j)} \int_t^r (r-s)^{j-1} ds
$$

$$
+ \frac{1}{\Gamma(\alpha)} \left| \int_0^1 (t-s)^{\alpha-1} s^{-\sigma} s^{\sigma} f(s, \overline{u}_s + y_s) ds \right.
$$

$$
\left. - \int_0^r (r-s)^{\alpha-1} s^{-\sigma} s^{\sigma} f(s, \overline{u}_s + y_s) ds \right|
$$

$$
\leq 2 \sum_{j=1}^{n-1} \lambda_j b^{\alpha-j} + l \sum_{j=1}^{n-1} \frac{a_j t^j}{\Gamma(j+1)}
$$

$$
+ l \sum_{j=1}^{n-1} \frac{a_j}{\Gamma(j+1)} \left[|r - t|^j + |r^j - t^j| \right]
$$

$$
+ \frac{NB(1-\sigma, \alpha)}{\Gamma(\alpha)} |t^{\alpha-\sigma} - r^{\alpha-\sigma}|
$$

$$
\leq 2 \sum_{j=1}^{n-1} \lambda_j b^{\alpha-j} + l \sum_{j=1}^{n-1} \frac{a_j b^j}{\Gamma(j+1)} + 2l \sum_{j=1}^{n-1} \frac{a_j}{\Gamma(j+1)}
$$

$$
+ l \sum_{j=1}^{n-1} \frac{a_j}{\Gamma(j+1)} |r - t|^j + \frac{2NB(1-\sigma, \alpha)}{\Gamma(\alpha)} b^{\alpha-\sigma}. \tag{6.26}
$$

Now, let us set $\Delta = \max\{3\Delta_1/\epsilon, 3\Delta_2/\epsilon, \Delta_3, \Delta_4, \cdots, \Delta_{n+1}\}$, where

$$
\Delta_1 = \sum_{j=1}^{n-1} \left(2\lambda_j b^{\alpha-j} + \frac{la_j b^j}{\Gamma(j+1)} + \frac{2la_j}{\Gamma(j+1)} \right),
$$

$$
\Delta_2 = \frac{2N}{\Gamma(\alpha)} B(1-\sigma, \alpha) b^{\alpha-\sigma}, \tag{6.27}
$$

$$
\Delta_{j+2} = l \sum_{j=1}^{n-1} \frac{a_j}{\Gamma(j+1)}, \quad j = 1, 2, \cdots, n-1.
$$

We have the following cases [Babakhani and Baleanu (2011)].

Case 1. If $0 < |r - t| < 1 < b$, then we choose $\delta = \epsilon/(3(n-1))$. As a result,

$$|Fu(t) - Fu(r)| \le \frac{\epsilon}{3}\Delta + \frac{\epsilon}{3}\Delta + \Delta \sum_{j=1}^{n-1} |r - t|^j$$

$$\le \frac{\epsilon}{3}\Delta + \frac{\epsilon}{3}\Delta + (n-1)\Delta\delta < \Delta\epsilon. \qquad (6.28)$$

Case 2. If $0 < 1 \le |r - t| < b$, then we choose $\delta = \{\epsilon/(3(n-1))\}^{1/(n-1)}$. Thus,

$$|Fu(t) - Fu(r)| \le \frac{\epsilon}{3}\Delta + \frac{\epsilon}{3}\Delta + \Delta \sum_{j=1}^{n-1} |r - t|^j$$

$$\le \frac{\epsilon}{3}\Delta + \frac{\epsilon}{3}\Delta + (n-1)\Delta\delta^{n-1} < \Delta\epsilon. \qquad (6.29)$$

As a result we proved that $F(G)$ is equicontinuous. We notice that the Arzelà-Ascoli Theorem implies that $\overline{F(G)}$ is compact.

By using the above results we conclude that $F : K \to K$ is completely continuous.

Theorem 6.3. ([Babakhani and Baleanu (2011)]) *Assume that, in addition to the hypotheses (6.1) – (6.3), there exists $\sigma \in (0, 1)$ such that $0 < \sigma < \alpha \in (n-1, n)$, $n \in \mathbb{N}$, and $t^\sigma f(t, x_t)$ is a continuous function on $[0, b]$. Then, (6.1) has at least one positive solution $x^* \in K^*$, fulfilling $\| x^* \| \le \max\{\| \omega \|, h\}$, where $h = 2 \wedge /(1 - \wedge) + 1$ and $\wedge$ is a positive constant to be specified in the proof of the theorem.*

Proof. *([Babakhani and Baleanu (2011)])* The operator $F : K \to K$ is continuous and it is completely continuous by using the Theorem 6.2. Below we prove that there exists an open set $U \subseteq K$, with $z \ne \gamma F(z)$ for $\gamma \in (0, 1)$ and $z \in \partial U$.

Let $z \in K$ be any solution of $z = \gamma F(z)$, $\gamma \in (0, 1)$. From Theorem 6.2, since $F : K \to K$ is continuous and it is completely continuous, we obtain

$$z(t) = \gamma Fz(t) = \gamma\{\lambda(t) + \mathcal{L}(I)z(t) + I^\alpha f(t, \overline{z}_t + y_t)\}, \quad t \in [0, b]$$

$$\le \sum_{j=1}^{n-1} \lambda_j b^j + \sum_{j=1}^{n-1} a_j |I^j z(t)| + |I^\alpha f(t, \overline{z}_t + y_t)| \qquad (6.30)$$

$$\leq \sum_{j=1}^{n-1} \lambda_j b^j + \| z \| \sum_{j=1}^{n-1} \frac{a_j}{\Gamma(j+1)} b^j$$

$$+ \frac{1}{\Gamma(\alpha)} \left| \int_0^t (t-s)^{\alpha-1} s^{-\sigma} s^\sigma f(s, \overline{z}_s + y_s) ds \right|.$$

The existence of a positive constant N such that $\| s^\sigma f(s, \overline{z}_s + y_s) \| \leq N$, as $s^\sigma f(s, \overline{z}_s + y_s)$ is continuous on $[0, b]$, implies that

$$z(t) \leq \sum_{j=1}^{n-1} \lambda_j b^j + \| z \| \sum_{j=1}^{n-1} \frac{a_j}{\Gamma(j+1)} b^j + \frac{N}{\Gamma(\alpha)} B(1-\sigma, \alpha). \quad (6.31)$$

Let us set $\wedge = \max\{\wedge_1, \wedge_2, \wedge_3\}$, where

$$\wedge_1 = \sum_{j=1}^{n-1} \lambda_j b^j, \quad \wedge_2 = \sum_{j=1}^{n-1} \frac{a_j}{\Gamma(j+1)} b^j, \quad \wedge_3 = \frac{N}{\Gamma(\alpha)} B(1-\sigma, \alpha). \ (6.32)$$

We notice that (6.24) lead us to

$$\| z \| \leq \wedge_1 + \| z \| \wedge_2 + \wedge_3 \leq 2\wedge + \| z \| A. \quad (6.33)$$

Thus, we obtain that $\| z \| (1-\wedge) \leq 2\wedge$. Thus, any solution $z = \gamma F(z)$ fulfills $\| z \| \neq h$. Let $U = \{z \in K : \| z \| < h\}$. Then, F has a fixed point $z \in \overline{U}$. From Theorem 6.1, (6.1) under the conditions of (6.3) admits a positive solution $x^\star \in K$ fulfilling $\| x^\star \| \leq \max\{\| \omega \|, h\}$.

6.2 Existence and Uniqueness for the Solution

In the following, we present the conditions on f and $a_1, a_2, \ldots, a_n$, which render a unique positive solution to (6.1).

Theorem 6.4. ([Babakhani and Baleanu (2011)]) *Let $f : (0, 1) \times [0, \infty) \to [0, \infty)$ be continuous and $\lim_{t \to 0} f(t, \cdot) = +\infty$. Suppose that there exists $\sigma \in (0, 1)$ so that $0 < \sigma < \alpha \in (n-1, n)$, $n \in \mathbb{N}$, and $t^\sigma f(t, x_t)$ is a continuous function on $[0, b]$. If further, the following conditions are satisfied (H1) $t^\sigma f(t, x_t)$ is Lipschitz with respect to the second variable with Lipschitz constant μ, that is,*

$$|t^\sigma f(t, x_t) - t^\sigma f(t, z_t)| \leq \mu \| \overline{u} - \overline{v} \|, \quad \forall u, v \in K, t \in (0, b], \quad (6.34)$$

(H2)

$$0 < \sum_{j=1}^{n-1} \frac{a_j b^j}{\Gamma(j+1)} + \frac{\mu b^{\alpha-\sigma}}{\Gamma(\alpha)} B(1-\sigma, \alpha) < 1, \quad (6.35)$$

where $x(t) = u(t) + y(t)$ and $z(t) = v(t) + y(t)$; then (6.1) under the conditions of (6.3) possess a unique positive solution.

Proof. *([Babakhani and Baleanu (2011)])* We know that (6.1) is equivalent to (6.17). Hence, for $u, v \in K$ we obtain

$$|Fu(t) - Fv(t)| = |\mathcal{L}(I)u(t) - \mathcal{L}(I)v(r)|$$
$$+ |I^\alpha f(t, \overline{u}_t + y_t) - I^\alpha f(r, \overline{v}_r + y_r)|$$
$$\leq \| u - v \|_b \sum_{j=1}^{n-1} \frac{a_j b^j}{\Gamma(j+1)}$$
$$+ |I^\alpha (f(t, \overline{u}_t + y_t) - f(t, \overline{v}_t + y_t))|$$
$$= \| u - v \|_b \sum_{j=1}^{n-1} \frac{a_j b^j}{\Gamma(j+1)}$$
$$+ |I^\alpha (t^{-\sigma} t^\sigma f(t, \overline{u}_t + y_t) - t^{-\sigma} t^\sigma f(t, \overline{v}_t + y_t))|$$
$$\leq \| u - v \|_b \sum_{j=1}^{n-1} \frac{a_j b^j}{\Gamma(j+1)} \tag{6.36}$$
$$+ \frac{1}{\Gamma(\alpha)} \max_{0 \leq t \leq b} \int_0^t (t-s)^{\alpha-1} s^{-\sigma}$$
$$\times |s^\sigma f(t, \overline{u}_t + y_t) - s^\sigma f(t, \overline{v}_t + y_t))| ds$$
$$\leq \| u - v \|_b \sum_{j=1}^{n-1} \frac{a_j b^j}{\Gamma(j+1)}$$
$$+ \frac{\mu}{\Gamma(\alpha)} \max_{0 \leq t \leq b} \int_0^1 (t-s)^{\alpha-1} s^{-\sigma} ds$$
$$\leq \| u - v \|_b \sum_{j=1}^{n-1} \frac{a_j b^j}{\Gamma(j+1)} + \frac{\mu \| u - v \|_b}{\Gamma(\alpha)} \max_{0 \leq t \leq b} t^{\alpha-\sigma}$$
$$\times B(1 - \sigma, \alpha)$$
$$\leq \| u - v \|_b \left\{ \sum_{j=1}^{n-1} \frac{a_j b^j}{\Gamma(j+1)} + \frac{\mu b^{\alpha-\sigma}}{\Gamma(\alpha)} B(1 - \sigma, \alpha) \right\},$$

where F is given in (6.20). As a result F has a unique fixed point in K, which is the unique positive solution of (6.1).

Chapter 7

Stability of a Class of Discrete Fractional Nonautonomous Systems

We notice that the discrete fractional calculus is a new concept suggested in [Miller and Ross (1989)]. Since that pioneering work several results were reported on this subject (see for example [Atici and Eloe (2007, 2009a,b); Abdeljawad and Baleanu (2011a); Abdeljawad (2011); Jarad et al. (2012); Baleanu et al. (2006); Abdeljawad et al. (2013); Abdeljawad and Baleanu (2011b)] and the references therein). This new concept describes better in some cases the dynamics of complex or hypercomplex systems. Some of the recent reported applications can be see in [Wu and Baleanu (2014a); Wu et al. (2014); Wu and Baleanu (2014b)]. One of the key concept in this new field is the stability of fractional-order dynamic systems which was already investigated in [Chen et al. (1995); Lazarevic (2006); Deng et al. (2007); Merrikh-Bayat and Karimi-Ghartemani (2009); Zhang (2008); Momani and Hadid (2004); Li et al. (2010); Jarad et al. (2012)] and the references therein, while the stability of discrete dynamic systems was a the subject of many books and articles such as [Agarwal (2000)] and the references therein.

7.0.1 *Preliminaries*

We define the set $\mathbb{N}_a = \{a, a + 1, a + 2, \ldots\}$ and $_b\mathbb{N} = b, b - 1, b - 2, \ldots$. The left fractional sum of f, defined on $\mathbb{N}_a$, of order α is defined as ([Miller and Ross (1989); Atici and Eloe (2007, 2009a,b); Abdeljawad and Baleanu (2011a); Abdeljawad (2011); Jarad et al. (2012)])

$$\Delta_a^{-\alpha} f(t) = \frac{1}{\Gamma(\alpha)} \sum_{s=a}^{t-\alpha} (t - \sigma(s))^{(\alpha-1)} f(s), \qquad (7.1)$$

where $\alpha > 0$, $\sigma(s) = s + 1$, and $t^{(\alpha)} = \Gamma(t + 1)/\Gamma(t + 1 - \alpha)$.

The right fractional sum of f, defined on $_b\mathbb{N}$, of order α is defined as

$$\nabla_b^{-\alpha} f(t) = \frac{1}{\Gamma(\alpha)} \sum_{s=t+\alpha}^{b} (\rho(s) - t)^{(\alpha-1)} f(s), \qquad (7.2)$$

where $\rho(s) = s - 1$.

The Riemann left and right fractional differences of order α are defined as [Jarad et al. (2012)]

$$\Delta_a^\alpha f(t) = \Delta^n \Delta_a^{-(n-\alpha)} f(t)$$
$$= \Delta^n \left[\frac{1}{\Gamma(n-\alpha)} \sum_{s=a}^{t-n+\alpha} (t - \sigma(s))^{(n-\alpha-1)} f(s) \right], \qquad (7.3)$$

and

$$\nabla_b^\alpha f(t) = (-1)^n \nabla^n \nabla_b^{-(n-\alpha)} f(t)$$
$$= (-1)^n \nabla^n \left[\frac{1}{\Gamma(n-\alpha)} \sum_{s=t+n-\alpha}^{b} (\rho(s) - t)^{(n-\alpha-1)} f(s) \right],$$

respectively, where $n = [\alpha] + 1$. $\Delta_a^{-\alpha}$ maps functions defined on $\mathbb{N}_a$ to functions defined on $\mathbb{N}_{a+\alpha}$ and Δ_a^α maps functions defined on $\mathbb{N}_a$ to functions defined on $\mathbb{N}_{a+n-\alpha}$. We notice that $\nabla_b^{-\alpha}$ maps functions defined on $_b\mathbb{N}$ to functions defined on $_{b-\alpha}\mathbb{N}_{a+\alpha}$ and ∇_b^α maps functions defined on b_N to functions defined on $_b\mathbb{N}$. The next step is to define the left and the right Caputo fractional differences of order α of a function defined on $\mathbb{N}_a$ and $_b\mathbb{N}$ respectively, by,

$$^C\Delta_a^\alpha f(t) = \Delta_a^{-(n-\alpha)} \Delta^n f(t)$$
$$= \frac{1}{\Gamma(n-\alpha)} \sum_{s=a}^{t-n+\alpha} (t - \sigma(s))^{(n-\alpha-1)} \Delta^n f(s), \qquad (7.4)$$

$$^C\nabla_b^\alpha f(t) = (-1)^n \nabla_b^{-(n-\alpha)} \nabla^n f(t)$$
$$= \frac{(-1)^n}{\Gamma(n-\alpha)} \sum_{s=t+n-\alpha}^{b} (\rho(s) - t)^{(n-\alpha-1)} \nabla^n f(s),$$

where $n = [\alpha] + 1$ [Abdeljawad (2011)].

7.0.2   *The Lyapunov Method for Discrete Fractional Nonautonomous Systems*

In the following we present the extension of the method of the Lyapunov functions to the study of the stability of solutions of the following system [Jarad et al. (2012)]:

$$^C\Delta_{t_0}^\alpha x(t) = g(t + \alpha - 1, x(t + \alpha - 1)), \quad x(t_0) = x_0, \qquad (7.5)$$

where $t_0 = a + n_0 \in \mathbb{N}_a(n_0 \in \mathbb{N})$, $t \in \mathbb{N}_{n_0}$, $a = \alpha - 1$, $g : \mathbb{N}_a \times \mathbb{R}^n \to \mathbb{R}^n$ is continuous, and $0 < \alpha \le 1$. It is clear that the Lyapunov function V of the system (7.5) depends on t and x. Let us consider $g(t, 0) = 0$, for all $t \in \mathbb{N}_a$ so that the system (7.5) has the trivial solution.

The basic definitions required in study of the stability properties of (7.5) are presented below.

Definition 7.1. [Jarad et al. (2012)] The trivial solution $x(t) = 0$ of (7.5) is said to be

(i) stable, if for each $\epsilon > 0$ and $t_0 \in \mathbb{N}_a$, there exists a $\delta = \delta(\epsilon, t_0) > 0$ such that for any solution $x(t) = x(t, t_0, x_0)$ with $\| x_0 \| < \delta$ one has $\| x(t) \| < \epsilon$, for all $t \in \mathbb{N}_{t_0} \subseteq \mathbb{N}_a$,

(ii) uniformly stable if it is stable and δ depends solely on ϵ,

(iii) asymptotically stable if it is stable and for all $t_0 \in \mathbb{N}_a$ there exists $\delta = \delta(t_0) > 0$ if $\| x_0 \| < \delta$ implies that $\lim_{t \to \infty} x(t, t_0, x_0) = 0$,

(iv) uniformly asymptotically stable if it is uniformly stable and, for each $\epsilon > 0$, there exists $T = T(\epsilon) \in \mathbb{N}_0$ and $\delta_0 > 0$ such that $\| x_0 \| < \delta_0$ implies $\| x(t) \| < \epsilon$ for all $t \in \mathbb{N}_{t_0 - \tau}$ and for all $t_0 \in \mathbb{N}_a$,

(v) globally asymptotically stable if it is asymptotically stable for all $x_0 \in \mathbb{R}^n$,

(vi) globally uniformly asymptotically stable if it is uniformly asymptotically stable for all $x_0 \in \mathbb{R}^n$.

Definition 7.2. [Jarad et al. (2012)] A function $\phi(r)$ is said to belong to the class $\mathcal{K}$ if and only if $\phi \in C[[0, \rho), \mathbb{R}]$, $\phi(0) = 0$, and $\phi(r)$ is strictly monotonically increasing in r. If $\psi : R_+ \to R_+ \psi \in \mathcal{K}$, and $\lim_{r \to \infty}(r) - \infty$, then ϕ is said to belong to class $\mathcal{KR}$.

Definition 7.3. [Jarad et al. (2012)] A real valued function $V(t, x)$ defined on $\mathbb{N}_a \times S_\rho$, where $S_\rho = \{x \in \mathbb{R}^n : \| x \| \le \rho\}$, is said to be positive definite if and only if $V(t, 0) = 0$ for all $t \in \mathbb{N}_a$ and there exists $\phi(r) \in \mathcal{K}$ such that $\phi(r) \le V(t, x), \| x \| = r, (t, x) \in \mathbb{N}_a \times S_\rho$.

Definition 7.4. [Jarad et al. (2012)] A real valued function $V(t, x)$ defined on $\mathbb{N}_a \times S_\rho$, where $S_\rho = \{x \in \mathbb{R}^n : \| x \| \ge \rho\}$, is said to be decrescent if and only if $V(t, 0) = 0$ for all $t \in \mathbb{N}_a$ and there exists $\psi(r) \in \mathcal{K}$ such that $V(t, x) \le \varphi(r), \| x \| = r, (t, x) \in \mathbb{N}_a \times S_\rho$.

The next step is to state the need theorems to prove the stability of solutions for (7.5).

Theorem 7.1. ([Jarad et al. (2012)]) *If there exists a positive definite and decrescent scalar function $V(t,x) \in C[\mathbb{N}_a \times S_\rho, \mathbb{R}_+]$ such that $_c\Delta_{t_0}^\alpha V(t,x(t)) \leq 0$ for all $t_0 \in \mathbb{N}_a$ and $(t,x) \in \mathbb{N}_0 \times S_\rho$, then the trivial solution of (7.5) is uniformly stable.*

Proof. *([Jarad et al. (2012)])* Let $x(t) = x(t,t_0,x_0)$ be a solution of system (7.5).

Since $V(t,x)$ is positive definite and decrescent, there exist $\phi, \varphi \in \mathcal{K}$ such that $\phi(\| x \|) \leq V(t,x) \leq \varphi(\| x \|)$ for all $(t,x) \in \mathbb{N}_a \times S_\rho$.

For each $\epsilon > 0$, $0 < \epsilon < \rho$, we choose a $\delta = \delta(\epsilon)$ such that $\varphi(\delta) < \phi(\epsilon)$. For any solution $x(t)$ of (7.5) we have $\phi(\| x(t) \|) \leq V(t,x(t))$ with $\| x_0 \| < \delta(\epsilon)$. Since $\Delta_{t_0}^\alpha V(t,x(t)) \leq 0$, we have $V(t,x(t)) \leq V(t_0,x_0)$ for all $t \in \mathbb{N}_{t_0}$. Consequently,

$$\phi(\| x(t) \|) \leq V(t,x(t)) \leq V(t_0,x_0) \leq \varphi(\| x_0 \|) < \varphi(\delta) < \phi(\epsilon), \quad (7.6)$$

and thus $\| x(t) \| < \epsilon$ for all $t \in \mathbb{N}_{t_0}$. $\square$

Theorem 7.2. ([Jarad et al. (2012)]) *If there exists a positive definite and decrescent scalar function $V(t,x) \in C[\mathbb{N}_a \times S_\rho, \mathbb{R}_+]$ such that*

$$^C\Delta_{t_0}^\alpha V(t,x(t)) \leq -\psi(\| x(t + \alpha - 1) \|), \quad \forall t_0 \in \mathbb{N}_a, \ (t,x) \in \mathbb{N}_0 \times S_\rho, \quad (7.7)$$

where $\psi \in \mathcal{K}$, then the trivial solution of (3.1) is uniformly asymptotically stable.

Proof. *([Jarad et al. (2012)])* Taking into account that all the conditions of Theorem 7.1 are fulfilled, the trivial solution of (7.5) is uniformly stable.

Let $0 < \epsilon < \rho$ and $\delta = \delta(\epsilon)$ correspond to uniform stability. Choose a fixed $\epsilon_0 \leq \rho$ and $\delta_0 = \delta(\epsilon_0) > 0$. Next, choose $\| x_0 \| < \delta_0$ and $T(\epsilon)$ large enough such that $(T + a)^{(\alpha)} \geq (\phi(\delta_0)/\psi(\delta(\epsilon)))\Gamma(\alpha + 1)$. A large T can be chosen since $\lim_{T \to \infty}(\Gamma T + \alpha)/\Gamma(T)) = \infty$.

Now, we claim that $\| +(t,t_0,x_0) \| < \delta(\epsilon)$ for all $t \in [t_0, t_0 + T] \cap \mathbb{N}_{t_0}$. If this is not true, due to (7.7), we obtain

$$V(t,x(t,t_0,x_0)) \leq V(t_0,x_0)$$

$$- \frac{1}{\Gamma(\alpha)} \sum_{s=t_0+1-\alpha}^{t-\alpha} (t - \sigma(s))^{(\alpha-1)} \psi(\| x(s + \alpha - 1) \|)$$

$$\leq \phi(\| x_0 \|) - \frac{\psi(\delta)}{\Gamma(\alpha)} \sum_{s=n_0}^{t-\alpha} (t - \sigma(s))^{(\alpha-1)} \quad (7.8)$$

$$\leq \phi(\delta_0) - \frac{\psi(\delta)}{\Gamma(\alpha + 1)} (t - n_0)^{(a)}.$$

Substituting $t = t_0 + T$, we obtain

$$0 < \psi(\delta(\epsilon)) \leq V(t_0 + T, x(t_0 + T, t_0, x_0))$$
$$\leq \phi(\delta_0) - \frac{\psi(\delta)}{\Gamma(\alpha + 1)}(T + t_0 - n_0)^{(\alpha)} \leq 0, \qquad (7.9)$$

which leads us to a contradiction.

Thus, there exists a $t \in [t_0, t_0 + T]$ such that $\| x(t) \| < \delta(\epsilon)$. But in this case, since the trivial solution is uniformly stable and t is arbitrary, $\| x(t) \| < \epsilon$ for all $t \geq t_0 + T$ whenever $\| x_0 \| < \delta_0$.

Theorem 7.3. ([Jarad et al. (2012)]) *If there exists a function $V(t, x) \in \mathcal{C}[\mathbb{N}_a \times \mathbb{R}^n, \mathbb{R}_+]$ such that*

$$\phi(\| x(t) \|) \leq V(t, x) \leq \varphi(\| x(t) \|) \quad \forall (t, x) \in \mathbb{N}_a \times \mathbb{R}^n,$$
$$(7.10)$$
$${}^{C}\Delta_{t_0}^{\alpha} V(t, x(t)) \leq -\psi(\| x(t + \alpha - 1) \|) \quad \forall t_0 \in \mathbb{N}_a, \ (t, x) \in \mathbb{N}_0 \times \mathbb{R}^n,$$

where ϕ, φ, and $\psi \in \mathcal{KR}$ hold for all $(t, x) \in \mathbb{N}_a \times \mathbb{R}^n$, then the trivial solution of (7.5) is globally uniformly asymptotically stable.

Proof. *([Jarad et al. (2012)])* We notice that the conditions of Theorem 7.2 are satisfied,therefore the trivial solution of (7.5) is uniformly asymptotically stable.

Now, we prove that the domain of attraction of $x = 0$ is all of $\mathbb{R}^n$. Knowing that $\lim_{r \to \infty} \phi(r) = \infty, \delta_0$ in the proof of Theorem 7.3 may be chosen arbitrary large and ϵ can be chosen such that it fulfils $\varphi(\delta_0) < \phi(\epsilon)$.

As a result the globally uniformly asymptotic stability of $x = 0$ is reported. $\square$

Chapter 8

Mittag-Leffler Stability of Fractional Nonlinear Systems with Delay

The dynamics of time-delay system is an important subject from both theoretical applied point of views [Richard (2003)]. We recall that the time delay appears naturally in various technical systems, e.g., electric, chemical, pneumatic, and hydraulic networks, long transmission lines and many others. Various results were reported on this matter, with particular emphasis on the application of Lyapunov's second method (see for example references [Chen et al. (1995); Lee and Dianat (1981)] and the references therein).

Mittag-Leffler stability is a relatively new concept related deeply to fractional calculus [Li et al. (2010)]. Seminal works on the applicability of $Lyapunov's$ second method in the case of time delay can be seen in [Razumikhin (1956); Krasovski (1956)].

As it is known, various types of the Lyapunov functions were proposed for the stability analysis of delay systems, see for example the pioneering works of Razumikhin [Razumikhin (1956)] and Krasovski [Krasovski (1956)]. We notice that Razumikhin [Razumikhin (1956)] used the Lyapunov-type functions $V(x(t))$ depending on the current value $x(t)$ of the solution, while Krasovski [Krasovski (1956)] proposed to use functionals $V(x_t)$ depending on the whole solution segment x_t, that is, the true state of the delay system, see [Hale and Verduyn Lunel (1993)] and the references therein for a plethora of details.

The readers can find more details about the stability of fractional-order linear time delay systems in [Lazarevic (2006); Zhang (2008); Sadati et al. (2010)] and the references therein. Regarding the Lyapunov's second method, some results in the field of stability of fractional order nonlinear systems without delay were reported in [Momani and Hadid (2004); Sabatier (2008); Li et al. (2010)].

The Mittag-Leffler function defined as

$$E_\alpha(z) = \sum_{k=0}^{\infty} \frac{z^k}{\Gamma(k\alpha + 1)}, \tag{8.1}$$

where $\alpha > 0$ and $z \in C$ looks like the queen of the fractional calculus. We notice that the important Mittag-Leffler function with two parameters has the following form:

$$E_{\alpha,\beta}(z) = \sum_{k=0}^{\infty} \frac{z^k}{\Gamma(k\alpha + \beta)}, \tag{8.2}$$

where $\alpha > 0$, $\beta > 0$, and $z \in \mathbb{C}$. By inspection, for $\beta = 1$ we get $E_{\alpha,1}(z) = E_\alpha(z)$. In addition we mention the following property $E_{1,1}(z) = e^z$. We consider $\mathcal{C}([a,b], \mathbb{R}^n)$ be the set of continuous functions mapping the interval $[a,b]$ to $\mathbb{R}^n$. If we would like to identify a *maximum time delay* r of a given system, thus, we are dealing with the set of continuous function mapping $[-r,0]$ to R^n, denoted by $\mathcal{C} = \mathcal{C}([-r,0], \mathbb{R}^n)$. Therefore, for any $A > 0$ and any continuous function of time $\psi \in \mathcal{C}([t_0-r, t_0+A], \mathbb{R}^n), t_0 \le t \le t_0+A$, let $\psi_t(\theta) \in \mathcal{C}$ be a segment of function ψ such that $\psi_t(\theta) = \psi(t+\theta), -r \le \theta \le 0$.

The next step is to discuss the Caputo fractional nonlinear time-delay system

$$_{t_0}C_t^q x(t) = f(t, x_t), \tag{8.3}$$

where $x(t) \in \mathbb{R}^n, 0 < \alpha \le 1$, and $f : \mathbb{R} \times \mathcal{C} \to \mathbb{R}^n$.

We notice that (8.3) indicates the Caputo derivatives of the state variable x on $[t_0, t]$ and $x(\xi)$ for $t - r \le \xi \le t$. In order to find the future evolution of the state, we have to specify the initial state variables $x(t)$ in a time interval of length r, namely, from $t_0 - r$ to t_0, in other words we have

$$x_{t_0} = \varphi, \tag{8.4}$$

where $\phi \in \mathcal{C}$ is provided. We conclude that we have $x(t_0 + \theta) = \phi(\theta), -r \le \theta \le 0$.

The Euclidean norm for vectors denoted by $\| \cdot \|$ will be used in the following. Also, we notice that the space of continuous initial functions $\mathcal{C}([-r,0], \mathbb{R}^n)$ is provided with the supremum norm [Sadati et al. (2010)],

$$\| \varphi \|_\infty = \max_{\theta \in [-r,0]} \| \varphi(\theta) \| . \tag{8.5}$$

Definition 8.1. [Sadati et al. (2010)] Let $V(t, \phi)$ be differentiable, and let $x_t(\tau, \varphi)$ be the solution of (8.3) at time t with initial condition $x_\tau = \varphi$.

Then, we calculate the Riemann-Liouville and the Caputo derivatives of $V(t, x_t)$ with respect to t and evaluate it at $t = \tau$ as follows, respectively,

$$
\begin{aligned}
{}_{t_0}D_t^q V(\tau, \varphi) &= \left. {}_{t_0}D_t^q V(t, x_t(\tau, \varphi)) \right|_{t=\tau, x_t=\varphi} \\
&= \left. \frac{1}{\Gamma(1-q)} \frac{d}{dt} \left(\int_{t_0}^t \frac{V(s, x_s)}{(t-s)^q} ds \right) \right|_{t=\tau, x_t=\varphi}
\end{aligned}
\tag{8.6}
$$

$$
\begin{aligned}
{}_{t_0}C_t^q V(\tau, \varphi) &= \left. {}_{t_0}C_t^q V(t, x_t(\tau, \varphi)) \right|_{t=\tau, x_t=\varphi} \\
&= \left. \frac{1}{\Gamma(1-q)} \int_{t_0}^t \frac{V'(s, x_s)}{(t-s)^q} ds \right|_{t=\tau, x_t=\varphi}
\end{aligned}
$$

where $0 < q \leq 1$.

Definition 8.2. [Sadati et al. (2010)] (exponential stability *[Kharitonov and Hinrichsen (2004)]*). The solution of (8.3) is said to be exponential stable if there exist $b > 0$ and $a \geq 0$ such that for every solution $x(t, \varphi), \varphi \in C([-r, 0], \mathbb{R}^n)$ the following exponential estimate holds

$$
\| x(t, \varphi) \| \leq a \| \varphi \|_\infty e^{-bt}.
\tag{8.7}
$$

Definition 8.3. [Sadati et al. (2010)] (Mittag-Leffler stability). The solution of (8.3) is said to be Mittag-Leffler stable if

$$
\| x(t, \varphi) \| \leq \{ m(\| \varphi \|_\infty) E_\alpha(-\lambda(t - t_0)^\alpha) \}^b,
\tag{8.8}
$$

where $a \in (0, 1)$, $\lambda \geq 0$, $b > 0$, $m(0) = 0$, $m(x) \geq 0$, and $m(x)$ is locally the Lipschitz on $x \in B \subset \mathbb{R}^n$ with the Lipschitz constant m_0.

Theorem 8.1. ([Sadati et al. (2010)]) *Suppose that α_1, α_2, and β are positive constants and $V, \omega : C([-r, 0], \mathbb{R}^n) \to \mathbb{R}$ are continuous functionals. If the following conditions are satisfied for all $\varphi \in C([-r, 0], \mathbb{R}^n)$:*
(1) $\alpha_1 \| \varphi(0) \|^2 \leq V(\varphi) \leq \alpha_2 \| \varphi \|_\infty^2$,
(2) $\beta V(\varphi) \leq \omega(\varphi)$,
(3) $V(x_t(\varphi))$ has fractional derivative of order α for all $\phi \in C([-r, 0], \mathbb{R}^n)$,
(4) ${}_{t_0}C_t^\gamma V(x_t(\varphi)) \leq -\omega(x_t(\varphi))$ for all $t \geq t_0$ and $0 < \gamma \leq 1$,
then

$$
\| x(t, \varphi) \| \leq \sqrt{\frac{\alpha_2}{\alpha_1}} \| \varphi \|_\infty (E_\gamma(-\beta t^\gamma))^{1/2}, \quad t \geq t_0,
\tag{8.9}
$$

where $\varphi \in C([-r, 0], \mathbb{R}^n)$. Thus, the solution of (8.3) is Mittag-Leffler stable.

Proof. [*Sadati et al. (2010)*] Given any $\varphi \in \mathcal{C}([-r, 0], \mathbb{R}^n)$, the condition (2) implies that

$$-\omega(\varphi) \leq -\beta V(\varphi). \tag{8.10}$$

From (8.10) and the condition (4), we have

$$_{t_0}C_t^\gamma V(x_t(\varphi)) + \beta V(x_t(\varphi)) \leq 0, \tag{8.11}$$

or

$$_{t_0}C_t^\gamma V(x_t(\varphi)) + \beta V(x_t(\varphi)) + M(t) = 0, \tag{8.12}$$

where $M(t) \geq 0$ for $t \geq 0$.

From (8.4), we obtain $V(x_{t_0}(\varphi)) = V(\phi)$. As a result, the solution of (8.12) together with initial condition $V(x_{t_0}(\varphi)) = V(\varphi)$ has the following form

$$V(x_t(\varphi)) = V(\varphi)E_\gamma(-\beta t^\gamma) - \int_{t_0}^t (t - \tau)^{\gamma-1} E_{\gamma,\gamma}(-\beta(t - \tau)^\gamma)M(\tau)d\tau$$
$$= V(\varphi)E_\gamma(-\beta t^\gamma) - M(t) * (t^{\gamma-1}E_{\gamma,\gamma}(-\beta t^\gamma)), \tag{8.13}$$

where $*$ is convolution operator.

Since both $t^{\gamma-1}$ and $E_{\gamma,\gamma}(-\beta t^\gamma)$ are nonnegative functions, we conclude that

$$V(x_t(\varphi)) \leq V(\varphi)E_\gamma(-\beta t^\gamma), \quad t \geq t_0. \tag{8.14}$$

Then, the above conditions (1) and (2) yield

$$\alpha_1 \parallel x(t, \varphi) \parallel^2 \leq V(x_t(\varphi))$$
$$\leq V(\varphi)E_\gamma(-\beta t^\gamma)$$
$$\leq \alpha_2 \parallel \varphi \parallel_\infty^2 E_\gamma(-\beta t^\gamma). \tag{8.15}$$

Comparing the left- and the right-hand sides, we obtain

$$\parallel x(t, \varphi) \parallel \leq \sqrt{\frac{\alpha_2}{\alpha_1}} \parallel \varphi \parallel_\infty (E_\gamma(-\beta t^\gamma))^{1/2}, \quad t \geq t_0, \tag{8.16}$$

where $\varphi \in \mathcal{C}([-\tau, 0], \mathbb{R}^n)$. $\square$

Lemma 8.1. ([Sadati et al. (2010)]) *Let* $\gamma \in (0, 1)$ *and* $V(0) \geq 0$, *then*

$$_{t_0}C_t^\gamma V(t) \leq _{t_0}D_t^\gamma V(t). \tag{8.17}$$

Proof. *[Sabatier (2008)]* We have $_{t_0}C_t^\gamma V(t) = {_{t_0}}D_t^\gamma V(t) - V(t_0)(t - t_0)^{-\gamma}/\Gamma(1 - \gamma)$. Because $\gamma \in (0,1)$ and $V(t_0) \geq 0$, we get $_{t_0}C_t^\gamma V(t) \leq {_{t_0}}D_t^\gamma V(t)$. $\square$

Theorem 8.2. ([Sadati et al. (2010)]) *If the assumptions in Theorem 8.1 are satisfied except replacing $_{t_0}C_t^\gamma$ by $_{t_0}D_t^\gamma$, then one has*

$$\| x(t, \varphi) \| \leq \sqrt{\frac{\alpha_2}{\alpha_1}} \| \varphi \|_\infty (E_\gamma(-\beta t^\gamma))^{1/2}, \tag{8.18}$$

where $t \geq t_0$, $\varphi \in \mathcal{C}([-\tau, 0], \mathbb{R}^n)$.

Proof. *[Sadati et al. (2010)]* By using the Lemma 8.1 and $V(x_t(\varphi)) \geq 0$ we obtain $_{t_0}^c D_t^\gamma V(x_t(\varphi)) \leq {_{t_0}}D_t^\gamma V(x_t(\varphi))$ which implies $_{t_0}C_t^\gamma V(x_t(\varphi)) \leq {_{t_0}}D_t^\gamma V(x_t(\varphi)) \leq -\omega(x_t(\varphi))$ for all $t \geq t_0$. Following the proof of Theorem 8.1 leads to

$$\| x(t, \varphi) \| \leq \sqrt{\frac{\alpha_2}{\alpha_1}} \| \varphi \|_\infty (E_\gamma(-\beta t^\gamma))^{1/2}, \tag{8.19}$$

where $t \geq t_0$, $\varphi \in \mathcal{C}([-\tau, 0], \mathbb{R}^n)$. $\square$

Corollary 8.1. ([Sadati et al. (2010)]) *For $\gamma = 1$ one has exponential stability* ([Kharitonov and Hinrichsen (2004)])

$$\| x(t, \varphi) \| \leq \sqrt{\frac{\alpha_2}{\alpha_1}} \| \varphi \|_\infty (E_1(-\beta t^\gamma))^{1/2}$$

$$= \sqrt{\frac{\alpha_2}{\alpha_1}} \| \varphi \|_\infty e^{(-(1/2)\beta t)}, \tag{8.20}$$

where $t \geq t_0$, $\varphi \in \mathcal{C}([-\tau, 0], \mathbb{R}^n)$.

Chapter 9

Razumikhin Stability for Fractional Systems in the Presence of Delay

As it well known, the Razumikhin stability theory is extensively utilized in the literature to prove the stability of time-delay systems [Hale and Verduyn Lunel (1993); Kequin et al. (2003)], mainly because the building of Lyapunov-Krasovskii functional seems more difficult than that of Lyapunov-Razumikhin function.

For more details, the reader can find some seminal papers regarding the stability of fractional-order linear time-delay systems in [Lazarevic (2006); Deng et al. (2007); Merrikh-Bayat and Karimi-Ghartemani (2009); Zhang (2008); Momani and Hadid (2004); Li et al. (2010); Hale and Verduyn Lunel (1993); Kequin et al. (2003)] and the references therein. Below we consider the following problem [Baleanu et al. (2010e)]

$$_{t_0}C_t^q x(t) = f(t, x_t), \tag{9.1}$$

where x(t) denotes a vector in R^n, $f : R \times C \to R^n$ and $0 < q \leq 1$.

We recall that an effective method for determining the stability of a time-delay system is the Lyapunov method. Having in mind that in a time-delay system the "state" at time t depends on the value of $x(t)$ in the interval $[t - r, t]$, that is, x_t, it is natural to expect that, for a time-delay system, the corresponding Lyapunov function be a functional $V(t, x_t)$ depending on x_t, which encapsulates the deviation of x_t apart the trivial solution [Baleanu et al. (2010e)].

Let $V(t, \phi)$ be differentiable, and let $x_t(\tau, \varphi)$ denoting the solution of (9.1) at time t equipped with the initial condition $x_\tau = \phi$. The next step is to evaluate the Caputo derivatives of $V(t, x_t)$ with respect to t and calculate it at $t = \tau$ as [Baleanu et al. (2010e)]

$$_{t_0}D_t^q V(\tau, \phi) = \left. _{t_0}D_t^q V(t, x_t(\tau, \phi)) \right|_{t=\tau, x_t=\phi}$$

$$= \frac{1}{\Gamma(1-q)} \frac{d}{dt} \left(\int_{t_0}^t \frac{V(s, x_s)}{(t-s)^q} ds \right) \Bigg|_{t=\tau, x_t=\phi}, \qquad (9.2)$$

$$_{t_0}C_t^q V(\tau, \phi) = \, _{t_0}C_t^q V(t, x_t(\tau, \phi)) \Big|_{t=\tau, x_t=\phi}$$

$$= \frac{1}{\Gamma(1-q)} \int_{t_0}^t \frac{V'(s, x_s)}{(t-s)^q} ds \Bigg|_{t=\tau, x_t=\phi}, \qquad (9.3)$$

where $0 < q \leq 1$.

Theorem 9.1. ([Baleanu et al. (2010e)]) *Suppose that $f : \mathbb{R} \times \mathcal{C} \to \mathbb{R}^n$ in (3.1) maps $\mathbb{R} \times$ (bounded sets in $\mathcal{C}$) into bounded sets in $\mathbb{R}^n$, and $\alpha_1, \alpha_2, \alpha_3 : \overline{\mathbb{R}}_+ \to \overline{\mathbb{R}}_+$ are continuous nondecreasing functions, $\alpha_1(s), \alpha_2(s)$ are positive for $s > 0$, and $\alpha_1(0) = \alpha_2(0) = 0$, α_2 strictly increasing. If there exists a continuously differentiable function $V : \mathbb{R} \times \mathbb{R}^n \to \mathbb{R}$ such that*

$$\alpha_1(\| x \|) \leq V(t, x) \leq \alpha_2(\| x \|), \quad \text{for } t \in \mathbb{R}, \, x \in \mathbb{R}^n, \qquad (9.4)$$

and the Caputo fractional derivative of V along the solution $x(t)$ of (9.1) satisfies

$$_{t_0}C_t^\gamma V(t, x(t))) \leq -\alpha_3(\| x(t) \|)$$
$$\text{whenever } \mathrm{V}(t + \theta, x(t + \theta)) \leq \mathrm{V}(t, x(t)) \qquad (9.5)$$

for $0 < q \leq 1$ and $\theta \in [-r, 0]$, then system (9.1) is uniformly stable.

If, in addition, $\alpha_3(s) > 0$ for $s > 0$ and there exists a continuous nondecreasing function $p(s) > s$ for $s > 0$ such that condition (9.5) is strengthened to

$$_{t_0}C_t^q \mathrm{V}(t, x(t)) \leq -\alpha_3(\| x(t) \|),$$
$$\text{if } \mathrm{V}(t + \theta, x(t + \theta)) \leq p(\mathrm{V}(t, x(t))) \qquad (9.6)$$

for $0 < q \leq 1$ and $\theta \in [-r, 0]$, then system (9.1) is uniformly asymptotically stable. If, in addition $\lim_{s \to \infty} \alpha_1(s) = \infty$, then system (9.1) is globally uniformly asymptotically stable.

We recall that the integer-order version of this theorem can be seen in [Hale and Verduyn Lunel (1993); Kequin et al. (2003)].

Proof. *([Baleanu et al. (2010e)])* The first step in proving the uniform stability is to consider for any given $\epsilon > 0$, the following $0 < \alpha_2(\delta) < \alpha_1(\epsilon)$. Thus, for any given t_0 and ϕ, such that $\| \phi \| < \delta$, we obtain $V(t_0 + \theta, \phi(\theta)) \leq \alpha_2(\delta) < \alpha_1(\epsilon)$ for $\theta \in [-r, 0]$.

We consider x be solution of (9.1) equipped with initial condition $x_{t_0} = \phi$. By using (9.5), as t increases, whenever $V(t, x(t)) = \alpha_2(\delta)$ and $V(t +$

$\theta, x(t + \theta)) \leq \alpha_2(\delta)$ for $\theta \in [-r, 0]$, $_{t_0}C_t^q V(t, x(t)) \leq 0$, therefore, by (9.4), $V(t, x(t)) \leq V(t_0, x(t_0)) \leq \alpha_2(\delta)$ for all $t \geq t_0$. We recall that the continuity of $V(t, x(t))$ implies that it is therefore impossible for $V(t, x(t))$ to be bigger than $\alpha_2(\delta)$. Thus, we obtain $V(t, x(t)) \leq \alpha_2(\delta) < \alpha_1(\epsilon)$ for $tt_0 - r$, and we notice that this implies that $\parallel x(t) \parallel \leq \epsilon$ for $t \geq t_0 - r$.

To end the proof of the theorem, assume that $\delta > 0, H > 0$ are such that $\alpha_2(\delta) = \alpha_1(H)$. We recall that such numbers always exist by our hypotheses on α_1 and α_2. In fact, since $\alpha_2(0) = 0$ and $0 < \alpha_1(s) \leq \alpha_2(s)$ for $s > 0$, one can preassign H and then find a δ such that the desired relation is fulfilled. We mention that this remark and the consequences that follows shows the uniform asymptotic stability of $x = 0$ together with the fact that $x = 0$ is a globally uniformly asymptotically stable [Baleanu et al. (2010e)].

When $\alpha_2(\delta) = \alpha_1(H)$, the same way of reasoning as in the proof of uniform stability implies that $\parallel \phi \parallel \leq \delta$ which leas us to the conclusion that $\parallel x(t) \parallel \leq H$, $V(t, x(t)) < \alpha_2(\delta)$ for $t \geq t_0 - r$. Let $0 < \eta \leq H$ be arbitrary. Our aim is to prove that there is a number $\bar{t} = \bar{t}(\eta, \delta)$ such that for any $t_0 \geq 0$ and $\parallel \phi \parallel \leq \delta$ the solution $x(t)$ of (9.1) fulfills $\parallel x(t) \parallel \leq \eta, t \geq t_0 + t + r$. This result is true provided that that $V(t, x(t)) \leq \alpha_1(\eta)$, for $t \geq t_0 + \bar{t}$.

By using the properties of function $p(s)$, we conclude that there is a number $a > 0$ obeying $p(s) - s > a$ for $\alpha_1(\eta) \leq s \leq \alpha_2(\delta)$. Let N be the first positive integer fulfilling $\alpha_1(\eta) + Na \geq \alpha_2(\delta)$, and let $\gamma = \inf_{\alpha_2^{-1}(\alpha_1(\eta) \leq s \leq H)} \alpha_3(s)$ and $T = (N\alpha_2(\delta)\Gamma(1 + q)/\gamma)^{1/q}$.

The next step is to prove that $V(t, x(t)) \leq \alpha_1(\eta)$ for all $t > t_0 + T$ [Baleanu et al. (2010e)]. First, we show that $V(t, x(t)) \leq \alpha_1(\eta) + (N - 1)a$ for $t \geq t_0 + (\alpha_2(\delta)\Gamma(1 + q)/\gamma)^{1/q}$. If $\alpha_1(\eta) + (N - 1)a < V(t, x(t))$, for $t_0 \leq t \leq t_0 + (\alpha_2(\delta)\Gamma(q + 1)/\Gamma)^{1/q}$, then, since $V(t, x(t)) \leq \alpha_2(\delta)$ for all $t \geq t_0 - r$, it follows that [Baleanu et al. (2010e)]

$$p(V(t, x(t))) > V(t, x(t)) + a \geq \alpha_1(\eta) + Na \geq \alpha_2(\delta)$$
$$\geq V(t + \theta, x(t + \theta)) \tag{9.7}$$

for $t_0 - r \leq t_0 + (\alpha_2(\delta)\Gamma(1 + q)/\gamma)^{1/q}$.

Making use of (9.6) we get that [Baleanu et al. (2010e)]

$$_{t_0}C_t^q V(t, x(t))) \leq -\alpha_3(\parallel x(t) \parallel) \leq -\gamma \tag{9.8}$$

for $t_0 \leq t \leq t_0 + (\alpha_2(\delta)\Gamma(1 + q)/\gamma)^{1/q}$. Thus,

$$_{t_0}C_t^q \left(V(t, x(t))) + \gamma \frac{(t - t_0)^q}{\Gamma(1 + q)} \right) \leq 0, \tag{9.9}$$

therefore for all

$$t_0 \leq t \leq t_0 + (\alpha_2(\delta)\Gamma(1+q)/\gamma)^{1/q} \tag{9.10}$$

we conclude that

$$V(t, x(t)) \leq V(t_0, x(t_0)) - \gamma \frac{(t-t_0)^q}{\Gamma(1+q)} \leq \alpha_2(\delta) - \gamma \frac{(t-t_0)^q}{\Gamma(1+q)} \tag{9.11}$$

on the same interval [Baleanu et al. (2010e)]. We recall that the positive property (9.3) of V implies that $V(t, x(t)) \leq \alpha_1(\eta) + (N-1)a$ at $t_1 = t_0 + (\alpha_2(\delta)\Gamma(1+q)/\gamma)^{1/q}$. This result implies that $V(t, x(t)) \leq \alpha_1(\eta) + (N-1)a$ for all $t \geq t_0 + (\alpha_2(\delta)\Gamma(1+q)/\gamma)^{1/q}$, since $_{t_0}C_t^q V(t, x(t))$ is negative by using (9.6), and as a result $(d/dt)(V(t, x(t)))$ is negative when $V(t, x(t)) = \alpha_1(\eta) + (N-1)a$.

Let $\bar{t}_j = (j\alpha_2(\delta)\Gamma(1+q)/\gamma)^{1/q}, j = 1, \ldots, N, \bar{t}_0 = 0$, and assume that, for some integer $k \geq 1$, in the interval $\bar{t}_{k-1} - r \leq t - t_0 \leq \bar{t}_k$, we have

$$\alpha_1(\eta) + (N-k)a \leq V(t, x(t)) \leq \alpha_1(\eta) + (N-k+1)a. \tag{9.12}$$

Using the same way of thinking as before, we have

$$_{t_0}C_t^q V(t, x(t)) \leq -\gamma \tag{9.13}$$

for $\bar{t}_{k-1} \leq t - t_0 \leq \bar{t}_k$, and we have

$$_{t_0}C_t^q \left(V(t, x(t)) + \gamma \frac{(t-t_0)^q}{\Gamma(1+q)} \right) \leq 0, \tag{9.14}$$

$$V(t, x(t)) + \gamma \frac{(t-t_0)^q}{\Gamma(1+q)} \leq V(t_0 + \bar{t}_{k-1}, x(t_0 + \bar{t}_{k-1}))$$
$$+ \gamma \frac{t_{k-1}^{-q}}{\Gamma(1+q)}, \tag{9.15}$$

and we have

$$V(t, x(t)) \leq V(t_0 + \bar{t}_{k-1}, x(t_0 + \bar{t}_{k-1})) + \gamma \frac{t_{k-1}^{-q}}{\Gamma(1+q)} - \frac{(t-t_0)^q}{\Gamma(1+q)},$$

$$\leq \alpha_2(\delta) - \gamma \frac{(t-t_0)^q}{\Gamma(1+q)} + (k-1)\alpha_2(\delta) \tag{9.16}$$

$$= k\alpha_2(\delta) - \gamma \frac{(t-t_0)^q}{\Gamma(1+q)} \leq 0$$

if $t - t_0 \geq (k\Gamma(1+q)\alpha_2(\delta)/\gamma)^{1/q}$. Thus,

$$V\left(t_0 + \bar{t}_k, x\left(t_0 + \bar{t}_{k-1}\right)\right) \leq \alpha_1(\eta) + (N-k)a \tag{9.17}$$

and finally, $V(t, x(t)) \leq \alpha_1(\eta) + (N-k)a$ for $t \geq t_0 + \bar{t}_k$. Therefore we end the induction and we conclude that $V(t, x(t)) \leq \alpha_1(\eta)$ for all $t \geq$

$t_0 + (N\alpha_2(\delta)\Gamma(1+q)/\gamma)^{1/q}$. As a result the theorem is proved [Baleanu et al. (2010e)]. □

Lemma 9.1. ([Baleanu et al. (2010e)]) *Let* $q \in (0,1)$ *and* $V(t_0) \geq 0$, *then*

$$_{t_0}C_t^q V(t) \leq {}_{t_0}D_t^q V(t). \tag{9.18}$$

Proof. *([Baleanu et al. (2010e)])* We have $_{t_0}C_t^q V(t) = {}_{t_0}D_t^q V(t) - V(t_0)(t-t_0)^{-q}/\Gamma(1-q)$. Taking into account that $q \in (0,1)$ and $V(t_0) \geq 0$, we conclude that $_{t_0}C_t^q V(t) \leq {}_{t_0}D_t^q V(t)$. □

Theorem 9.2. ([Baleanu et al. (2010e)]) *Suppose that the assumptions in Theorem 9.1 are satisfied except replacing* $_{t_0}C_t^q$ *by* $_{t_0}D_t^q$, *then one has the same result for uniform stability, uniform asymptotic stability, and global uniform asymptotic stability.*

Proof. *([Baleanu et al. (2010e)])* By making use of 9.1 and $V(t, x(t)) \geq 0$ we have $_{t_0}C_t^q V(t, x(t)) \leq {}_{t_0}D_t^q V(t, x(t))$, which implies that $_{t_0}C_t^q V(x_t(\varphi)) \leq {}_{t_0}D_t^q V(x_t(\varphi)) \leq -\omega(x_t(\varphi))$ for all $t \geq t_0$. By utilizing the same proof of Theorem 9.1 yields uniform stability, uniform asymptotic stability, and global uniform asymptotic stability, respectively. □

Chapter 10

Controllability of Some Fractional Evolution Nonlocal Impulsive Quasilinear Delay Integro-Differential Systems

10.1 Preliminaries

As it is known the controllability represents one of the fundamental concepts in mathematical control theory and has a powerful role in deterministic and stochastic control systems [Ogata (2009); Nise (2010)]. We recall that the controllability has extensive industrial and biological applications [Ogata (2009); Nise (2010); Oustaloup (1991); Oustaloup et al. (2006); Luo and Chen (2013); Baleanu et al. (2012b); Monje et al. (2010); Caponetto et al. (2010)].

Differential equations with impulsive effects are used to investigate many real world applications e.g., pharmacokinetics, the radiation of electromagnetic waves, population dynamics, biological systems, the abrupt increase of glycerol in fed-batch culture, bio-technology, nano-electronics and many others (see for example [Tai and Wang (2009); Lakshmikantham et al. (1989)] and the references therein). The basic theory of impulsive differential equations can be found in [Lakshmikantham et al. (1989)].

We recall that Byszewski [Byszewski (1991a,b)] initiated the existence results to evolution equations with nonlocal conditions in Banach space. Deng [Deng (1993)] used the nonlocal condition $u(0) + h(u) = u_0$ to investigate the diffusion phenomenon of a small amount of gas in a transparent tube and proved that it gives better result than using the standard local Cauchy problem $u(0) = u_0$. We notice that the function h is considered as given in [Deng (1993); Debbouche and Baleanu (2011)]

$$h(u) = \sum_{k=1}^{p} c_k u(t_k), \tag{10.1}$$

where $c_k, k = 1, 2, ..., p$ are constants and $0 \leq t_1 < \cdots < t_p \leq a$.

Controllability of different types of dynamical systems was debated in many papers (see for example [Tai and Wang (2009); Subalakshmi and Balachandran (2009); Balachandran and Park (2009); Balachandran and Sakthivel (2001); Sakthivel et al. (2004); Sakthivel et al. (2005)] and the references therein). Besides, the controllability of impulsive functional differential inclusions in Banach spaces can be found in [Abada et.al. (2009); Benchohra (2004)] and the approximate controllability of impulsive differential equations with state-dependent delay was reported in [Sakthivel and Anandhi (2009)]. In [Jeong et al. (2007)] it was investigated the controllability for semilinear retarded control systems in Hilbert spaces. In [Tai and Wang (2009)] was reported the sufficient conditions for the controllability of fractional impulsive neutral functional integro-differential systems in a Banach space. The controllability of fractional integro-differential systems in Banach spaces was analyzed in [Balachandran and Park (2009)]. For the existence result of fractional evolution equations the readers can consult [Agarwal et al. (2010c,a)] and the references therein. We recall that some different types of mild solutions were reported in the literature. The first one was constructed in terms of a probability density function by El-Borai [El-Borai (2002)] and developed by Zhou et al. in [Zhou and Jiao (2010a); Wang and Zhou (2011)], The second one was reported in terms of an A-resolvent family and it was provided in [Araya and Lizama (2008); Mophou and N'Guerekata (2010)].

We recall a new concept called (α, u)-resolvent family [Debbouche and Baleanu (2011)], which was based on Araya-Lizama concepts [Araya and Lizama (2008)], and Hille-Phillips principles [Hille and Phillips (1957)].

Let E be the Banach space formed from $D(A)$ with the graph norm. Since $-A(t)$ is a closed operator, it follows that $-A(t)$ is in the set of bounded operators from E to X.

Definition 10.1. [Debbouche and Baleanu (2011)] Let $A(t, u)$ be a closed and linear operator with domain $D(A)$ defined on a Banach space X and $\alpha > 0$. Let $\rho[A(t, u)]$ be the resolvent set of $A(t, u)$. We call $A(t, u)$ the generator of an (α, u)-resolvent family if there exist ($\omega \geq 0$ and a strongly continuous function $R_{(\alpha, u)} : \mathbb{R}^2 \to L(X)$ such that $\{\lambda^\alpha : Re(\lambda) > \omega\} \subset$

$\rho(A)$ and for $0 \le s \le t \le \infty$,

$$(\lambda^\alpha I - A(s, u))^{-1} v = \int_0^\infty e^{-\lambda(t-s)} R_{(\alpha,u)}(t, s) v \, dt, \qquad (10.2)$$

where $Re(\lambda) > \omega$, $(u, v) \in X^2$. In this case, $R_{(\alpha,u)}(t, s)$ is called the (α, u)-resolvent family generated by $A(t, u)$.

Remark 10.1. [Debbouche and Baleanu (2011)]

(1) If we delete s and u, then (10.1) becomes the introduced concept in [Araya and Lizama (2008)].
(2) We conclude that (10.4)-(10.4) is well posed if and only if, $-A(t, u)$ is the generator of (α, u)-resolvent family.
(3) $R_{(\alpha,u)}(t, s)$ can be extracted from the evolution operator of the generator $-A(t, u)$.
(4) The (α, u)-resolvent family is similar to the evolution operator for non-autonomous differential equations in a Banach space.

In the following we consider Ω be a subset of X.

Definition 10.2. [Debbouche and Baleanu (2011)] *(Compare [Debbouche (2010); Tai and Wang (2009); Subalakshmi and Balachandran (2009)] with [Araya and Lizama (2008)]).* By a mild solution of (10.4) – (10.6) we mean a function $u \in PC(J : X)$ with values in Ω satisfying the integral equation

$$u_\mu(t) = R_{(\alpha,u)}(t, 0)u_0 - R_{(\alpha,u)}(t, 0)h(u)$$

$$+ \int_0^t R_{(\alpha,u)}(t, s) \left[(B\mu)(s) \right.$$

$$+ \phi \left(s, f(s, u(\beta(s))), \int_0^s g(s, \eta, u(\gamma(\eta))) d\eta \right) \right] ds$$

$$+ \sum_{0 < t_i < t} R_{(\alpha,u)}(t, t_i) I_i(u(t_i)), \quad t \in J, \qquad (10.3)$$

for all $u_0 \in X$ and admissible control $\mu \in L^2(L, U)$.

Below we assume the following conditions [Debbouche and Baleanu (2011)].

(A_1) The bounded linear operator $W : L^2(J, U) \to X$, namely

$$W\mu = \int_0^a R_{(\alpha,u)}(a, s) B\mu(s) ds$$

admits an induced inverse operator $\tilde{W}^{-1}$ which takes values in $L^2(J, U) \backslash \ker W$ and there exist positive constants M_1, M_2, fulfilling $\|B\| \le M_1$ and $\|\tilde{W}^{-1}\| \le M_2$.

$(A_2)\, h : PC(J : \Omega) \to Y$ is Lipschitz continuous in X and bounded in Y, namely, there exist constants $k_1 > 0$ and $k_2 > 0$ fulfilling

$$\|h(u)\|_Y \le k_1,$$
$$\|h(u) - h(v)\|_Y \le k_2 \max_{t \in J} \|u - v\|_{PC}, \quad u, v \in PC(J : X).$$

For the next conditions $(A_3) - (A_5)$ let Z be taken as both X and Y.

$(A_3)\, g : \wedge \times Z^k \to Z$ is continuous and we have constants $k_3 > 0$ and $k_4 > 0$ such that

$$\int_0^t \|g(t, s, u_1, ..., u_k) - g(t, s, v_1, ...v_k)\|_Z ds \le k_3 \sum_{q=1}^k \|u_q - v_q\|_Z,$$

where $u_q,\, v_q \in X$, $q = 1, 2, ..., k$,

$$k_4 = \max \left\{ \int_0^t \|g(t, s, 0, ...0)\|_Z ds : \quad (t, s) \in \wedge \right\}.$$

$(A_4)\, f : J \times Z^r \to Z$ is continuous and there exist constants $k_5 > 0$ and $k_6 > 0$ obeying

$$\|f(t, u_1, ..., u_r) - f(t, v_1, ...v_r)\|_Z \le k_5 \sum_{p=1}^r \|u_p - v_p\|_Z,$$

where $u_p,\, v_p \in X$, $p = 1, 2, ..., r$ and

$$k_6 = \max \|f(t, 0, ...0)\|_Z.$$

$(A_5)\, \phi : J \times Z^2 \to Z$ is continuous and there exist constants $k_7 > 0$ and $k_8 > 0$

$$\|\phi(t, u_1, u_2) - \phi(t, v_1, v_2)\|_Z \le k_7(\|u_1 - v_1\|_Z + \|u_2 - v_2\|_Z),$$

where $u_1,\, u_2, v_1, v_2 \in X$ and

$$k_8 = \max_{t \in J} \|\phi(t, 0, 0)\|_Z.$$

$(A_6)\, \beta_p, \gamma_q : J \to J$ are bijective absolutely continuous and there exist constants $c_p > 0$ and $b_q > 0$ such that $\beta_p'(t) \ge c_p$ and $\gamma_q'(t) \ge b_q$ respectively for $t \in J$, $p = 1, ..., r$ and $q = 1, ..., k$.

$(A_7)\, I_i : X \to X$ are continuous and there exist constants $I_i > 0$, $i = 1, 2, ..., m$ fulfilling

$$\|I_i(u) - I_i\|(v) \le I_i \|u - v\|, \quad u, v \in X.$$

Now, let us consider $M_0 = \max \|R_{(\alpha,u)}(t,s)\|_{B(Z)}$, $0 \leq s \leq t \leq a, u \in \Gamma$.
(A_8) There exist positive constant $\delta_1, \delta_2, \delta_3 \in (0, \delta/3]$ and $\lambda_i, \lambda_2, \lambda_3$, $\lambda_4 \in [0, \frac{1}{4}]$ in such a way that

$$\delta_1 = M_0\|u_0\| + M_0k_1,$$
$$\delta_2 = M_0M_1M_2[\|u_1\| + M_0\|u_0\| + M_0k_1 + M_0k_7\theta + M_0k_8a + M_0\xi]a,$$
$$\delta_3 = M_0k_7\theta + M_0k_8a + M_0\xi,$$

and

$$\lambda_1 = Ka_0\|u_0\| + k_1K_a + M_0k_2,$$
$$\lambda_2 = 2a^2KM_1M_2\{\|u_1\|_Y + M_0(\|u_0\|_Y + k_1 + k_7\theta + k_8a + \xi)\},$$
$$\lambda_3 = Ka(k_7\theta + k_8a) + M_0k_7\rho,$$
$$\lambda_4 = Ka\xi + M_0 + \sum_{i=1}^{m} l_i,$$

where $\rho = a[k_5(1/c_1 + ... + 1/c_r) + k_3(1/b_1 + ... + 1/b_k)]$, $\theta = \rho\delta + a(k_4 + k_6)$ and $\Sigma_{i=1}^{m}(I_i\delta + \|l_i(0)\|)$.

10.2   The Problem

In the following we analyze the fractional nonlocal impulsive integro-differential control system ,namely [Debbouche and Baleanu (2011)]

$$\frac{d^\alpha u(t)}{dt^\alpha} + A(t, u(t))u(t) = (B\mu)(t) \tag{10.4}$$

$$+ \phi\left(t, f(t, u(\beta(t))), \int_0^t g(t, s, u(\gamma(s)))ds\right),$$
$$u(0) + h(u) = u_0 \tag{10.5}$$
$$\Delta u(t_i) + I_i(u(t_i)). \tag{10.6}$$

Here $u(.)$ takes values in the Banach space $X, 0 < \alpha \leq 1, t \in [0,a], u_0 \in X, i = 1, 2, ..., m$ and $0 < t_1 < t_2 < \cdots < t_m < a$.

The closed linear operator $-A(t,.)$ is defined on a dense domain $D(A)$ in X into X such that $D(A)$ is independent of t. We suppose that $-A(t,.)$ generates an evolution operator in the Banach space X, the control function belongs to the spaces $L^2(J,U)$, a Banach space of admissible control functions with U as a Banach space, and $B : U \to X$ represents a bounded linear operator [Debbouche and Baleanu (2011)].

Besides, the functions $f : J \times X^\alpha \to X, g : \wedge \times X^k \to X, \phi : J \times X^2 \to X, h : PC(J, X) \to X, u(\beta) = (u(\beta_1), ..., u(\beta_r)), u(\gamma) = (u(\gamma_1), ..., u(\gamma_k)),$ and $\beta_p, \gamma_q : J \to J$ are given and $p = 1, 2, ..., r, q = 1, 2, ..., k$. Here we have $J = [0, a]$ and $\wedge = \{(t, s) : 0 \le s \le t \le a\}$. [Debbouche and Baleanu (2011)]

Let $PC(J, X)$ consist of functions u from J into X, with $u(t)$ being continuous at $t \neq t_i$ and left continuous at $t = t_i$ and the right limit $u(t_i^+)$ exists for $i = 1, 2, ..., m$. We recall that $PC(J, X)$ denotes a Banach space equipped with the norm $\|u\|_{PC} = sup_{t \in J} \|u(t)\|$, and let $\Delta u(t_i) = u(t_i^+) - u(t_i^-)$ constitute an impulsive condition [Debbouche and Baleanu (2011)].

10.3   A Controllability Result

Definition 10.3. [Debbouche and Baleanu (2011)] We say that the fractional system (10.4) – (10.6) is controllable on the interval J if for all $u_0, u_1 \in X$, there exists a control $\mu \in L^2(J, U)$, such that the mild solution $u(.)$ of (10.4) – (10.6) corresponding to μ, verifies: $u(0) + h(u) = u_0, \Delta u(t_i) = I_i(u(t_i)), i = 1, 2, ..., m$ and $u_\mu(a) = u_1$.

Lemma 10.1. [Debbouche and Baleanu (2011)] *Let* $R_{(\alpha,u)}(t, s)$ *be the* (α, u) *-resolvent family for the fractional problem (10.4) – (10.6). There exists a constant* $K > 0$ *such that*

$$\|R_{(\alpha,u)}(t, s)\omega - R_{(\alpha,v)}(t, s)\omega\| \le K \|\omega\|_Y \int_s^t \|u(\tau) - v(\tau)\| d\tau,$$

for every $u, v \in PC(J : X)$ *with values in* Γ *and every* $\omega \in Y$.

Proof. [*Debbouche and Baleanu (2011)*] Having in mind that the resolvent operator is similarly to the evolution operator for nonautonomous differential equations in a Banach space, we can use a similar technique as described in [Pazy (1993), Lemma 4.4, p. 202].

Let $S_\delta = \{u : u \in PC(J : X), u(0) + h(u) = u_0, \Delta u(t_i) = I_i(u(t_i)), \|u\| \le \delta\}$, for $t \in J, \delta > 0, u_0 \in X$ and $i = 1, ..., m$.

Lemma 10.2. [Debbouche and Baleanu (2011)] *We have*

$$\|\varphi(t)\| \le 0,$$

where

$$\varphi(t) = \int_0^t \left[f(s, u(\beta(s))) + \int_0^s g(s, \tau, u(\gamma)) d\tau \right] ds.$$

Proof. *[Debbouche and Baleanu (2011)]* We notice that we have

$$\|\varphi(t)\| \leq \int_0^t [\|f(s, u(\beta_1(s)), ...u(\beta_r(s))) - f(s, 0, ..., 0)$$
$$+ \|f(s, 0, ..., 0)\|$$
$$+ \int_0^t \|g(s, \tau, u(\gamma_1(\tau)), ...u(\gamma_k(\tau))) - g(s, \tau, 0, ..., 0)\|d\tau$$
$$+ \int_0^s \|g(s, \tau, 0, ..., 0)\|d\tau]ds.$$

Taking into account A_3, A_4 and A_6, we obtain

$$\|\varphi(t)\| \leq \int_0^t [k_5(\|u(\beta_1(s))\| + \cdots + \|u(\beta_r(s))\|) + k_6$$
$$+ k_3(\|u(\gamma_1(s))\| + \cdots + \|u(\gamma_k(s))\|) + k_4]ds$$
$$+ \int_0^t [k_5\|u(\beta_1(s))\|(\beta_1'(s)/c_1) + \cdots + \|u(\beta_r(s))\|(\beta_r'(s/c_r) + k_6$$
$$+ k_3\{\|u(\beta_1(s))\|(\beta_1'(s)/b_1) + \cdots + \|u(\beta_k(s))\|(\beta_k'(s) b_k)\} + k_4]ds$$
$$\leq \frac{k_5}{c_1} \int_{\beta_1(0)}^{\beta_1(t)} \|u(\tau)\|d\tau + \cdots + \frac{k_5}{c_r} \int_{\beta_r(0)}^{\beta_r(t)} \|u(\tau)\|d\tau + k_6 a$$
$$+ \frac{k_3}{b_1} \int_{\gamma_1(0)}^{\gamma_1(t)} \|u(\eta)\|d\eta + \cdots + \frac{k_3}{b_k} \int_{\gamma_k(0)}^{\gamma_k(t)} \|u(\eta) + k_4 a\|.$$

So, the desired result was proved. $\square$

Theorem 10.1. ([Debbouche and Baleanu (2011)]) *Suppose that the operator $-A(t, u)$ generates an (α, u)-resolvent family with $\|R_{(\alpha,u)}(t, s)\| < Me^{-\sigma(t-s)}$ for some constants $M, \sigma > 0$. If hypotheses $(A_1) - (A_8)$ are satisfied, then the fractional control integro-differential system (10.4) with nonlocal condition (10.5) and impulsive condition (10.6) is controllable on J.*

Proof. *[Debbouche and Baleanu (2011)]* Using hypothesis (A_1), for an arbitrary function $u(.)$, we define the control

$$\mu(t) = \tilde{W}^{-1}[u_1 - R_{(\alpha,u)}(a, u)u_0 + R_{(\alpha,u)}(a, u)h(u)$$
$$- \int_0^a R_{(\alpha,u)}(a, s)\phi(s, f(s, u(\beta(s))), \int_0^s g(s, \eta, u(\gamma(\eta)))d\eta)ds$$
$$- \sum_{i=1}^m R_{(\alpha,u)}(a, t_i)I_i(u(t_i))](t).$$

We define an operator $P : S_\delta \to S_\delta$ by

$$(Pu_\mu)(t) = R_{\alpha,u}(t,0)u_0 - R_{\alpha,u}(t,0)h(u) + \int_0^t R_{\alpha,u}(t,\eta)B\tilde{W}^{-1}$$
$$[u_1 - R_{(\alpha,u)}(a,0)u_0 + R_{(\alpha,u)}(a,0)h(u)$$
$$- \int_0^a R_{(\alpha,u)(a,s)\phi}(s,f(s,u(\beta(s))), \int_0^s g(s,tau,u(\gamma(\tau)))d\tau)ds$$
$$- \sum_{i=1}^m R(\alpha,u)(a,t_i)I_i(u(t_i))](\eta)d\eta$$
$$+ \int_0^a R_{(\alpha,u)(t,s)\phi}(s,f(s,u(\beta(s))), \int_0^s g(s,tau,u(\gamma(\tau)))d\tau)ds$$
$$+ \sum_{0<t_i<t} R(\alpha,u)(t,t_i)I_i(u(t_i)).$$

Using this controller we prove that the operator P admits a fixed point [Debbouche and Baleanu (2011)] which is a solution of the equation (10.2). Clearly $Pu_{(\mu)}(a) = u_1$, which shows that the control μ steers system (10.4) – (10.6) from the initial state u_0 to u_1 in time a, provided we can obtain a fixed point of the nonlinear operator P.

Now, our aim is to show that P maps $S\delta$ into itself.

$$\|(Pu_\mu)(t)\| \le \|R_{\alpha,u}(t,0)u_0\| + \|R_{\alpha,u}(t,0)h(u)\|$$
$$+ \int_0^t \|R_{\alpha,u}(t,\eta)\|\|B\tilde{W}^{-1}\|[\|u_1\| + \|R_{\alpha,u}(a,0)u_0\|$$
$$+ \|R_{\alpha,u}(a,0)h(u)\| + \int_0^a \|R_{\alpha,u}(a,s)\|$$
$$\times \{\|\phi(s,f(s,u(\beta(s))), \int_0^s g(s,\tau,u(\gamma(\tau)))d\tau) - \phi(s,0,0)\|$$
$$+ \|\phi(s,0,0)\|\}ds$$
$$+ \sum_{i=1}^m \|R_{(\alpha,u)}(a,t_i)\|\{\|I_i(u)(t_i) - I_i(0)\| + \|I_i(0)\|\}]d\eta$$
$$+ \int_0^t \|R_{(\alpha,u)}(t,s)\|$$
$$\times \{\|\phi(s,f(s,u(\beta(s))), \int_0^s g(s,\tau,u(\gamma(\tau)))d\tau) - \phi(s,0,0)\|$$
$$+ \|\phi(s,0,0)\|\}ds$$
$$+ \sum_{0<t_i<t} \|R_{(\alpha,u)}(t,t_i)\|\{\|I_i(u)(t_i) - I_i(0)\| + \|I_i(0)\|\}.$$

From A_1, A_2, A_5 and A_7, we conclude

$$\|Pu_\mu(t)\| \leq M_0\|u_0\| + M_0 k_1 + \int_0^t M_0 M_1 M_2 [\|u_1\| + M_0\|u_0\| + M_0 k_1$$
$$+ \int_0^a M_0 \left\{ k_7 \left(\|f(s, u(\beta(s)))\| + \|\int_0^s g(s, \tau, u(\gamma(\tau)))d\tau\| \right) \right.$$
$$+ k_8 \} \, ds + M_0 \sum_{i=1}^m (I_i \delta + \|I_i(0)\|)]d\eta$$
$$+ \int_0^t M_0 \left\{ k_7 \left(\|f(s, u(\beta(s)))\| + \|\int_0^s g(s, \tau, u(\gamma(\tau)))d\tau\| \right) \right.$$
$$+ k_8 \} \, ds + M_0 \sum_{i=1}^m (I_i \delta + \|I_i(0)\|).$$

By using the Lemma 10.2 and A_8, we obtain

$$\|Pu_\mu(t)\| \leq M_0\|u_0\| + M_0 k_1$$
$$+ M_0 M_1 M_2 \left[\|u_1\| + M_0\|M_0 k_1 + M_0 k_7 \theta + M_0 k_8 a + M_0 \xi\right] a$$
$$+ M_0 k_7 \theta + M_0 k_8 a + M_0 \xi.$$

It is easy to show that the assumption A_8, provides $\|Pu_\mu(t)\| \leq \delta$. Thus, P maps S_δ into itself.

For $u, v \in S_\delta$, we conclude that

$$\|Pu_\mu(t) - Pv_\mu(t)\| \leq I_1 + I_2 + I_3 + I_4,$$

where

$$I_1 = \|R_{(\alpha,u)}(t,0)u_0 - R_{(\alpha,v)}(t,0)u_0\|$$
$$+ \|R_{(\alpha,u)}(t,0)h(u) - R_{(\alpha,v)}(t,0)h(v)\|,$$
$$I_2 = \int_0^t \{\|R_{(\alpha,u)}(t,\eta)B\tilde{W}^{-1}[u_1$$
$$- R_{(\alpha,u)}(a,0)u_0 + R_{(\alpha,u)}(a,0)h(u)$$
$$- \int_0^a R_{(\alpha,u)}(a,s)\phi(s, f(s, u(\beta(s))), \int_0^s g(s, \tau, u(\gamma(\tau)))d\tau)ds$$
$$- \sum_{i=1}^m R_{(\alpha,u)}(a, t_i)I_i(u(t_i))]$$
$$- R_{(\alpha,u)}(t,\eta)B\tilde{W}^{-1} \left[u_1 - R_{(\alpha,u)}(a,0)u_0 - R_{(\alpha,v)}(a,0)h(v)\right]$$

$$- \int_0^a R_{(\alpha,v)}(a,s)\phi(s,f(s,v(\beta(s))), \int_0^s g(s,\tau,v(\gamma(\tau)))d\tau)ds$$

$$- \sum_{i=1}^m R_{(\alpha,v)}(a,t_i)I_i(v(t_i))]\|\}d\eta,$$

$$I_3 = \int_0^a \|R_{(\alpha,u)}(t,s)\phi(s,f(s,u(\beta(s))), \int_0^s g(s,\tau,u(\gamma(\tau)))d\tau)$$

$$- R_{(\alpha,v)}(t,s)\phi(s,f(s,v(\beta(s))), \int_0^s g(s,\tau,v(\gamma(\tau)))d\tau)\|ds$$

and

$$I_4 = \sum_{i=1}^m \|R_{(\alpha,u)}(t,t_i)I_i(u(t_i)) - R_{(\alpha,v)}(t,t_i)I_i(v(t_i))\|.$$

By using the Lemma 10.1 and A_2, we obtain

$$I_1 \leq \|R_{(\alpha,u)}(t,0)u_0 - R_{(\alpha,v)}(t,0)u_0\|$$
$$+ \|R_{(\alpha,u)}(t,0)h(u) - R_{(\alpha,v)}(t,0)h(u)\|$$
$$+ R_{(\alpha,v)}(t,0)h(u) - R_{(\alpha,v)}(t,0)h(v)\|$$
$$\leq \{Ka\|u_0\| + k_1 Ka + M_0 k_2\} \max_{\tau \in J} \|u(\tau) - v(\tau)\|.$$

Lemmas 10.1 together with 10.2, A_1, A_2, A_5 and A_8, give

$$I_2 \leq a^2 KM_1M_2\{\|2\max([u_1 - R_{(\alpha,u)}(a,0) + R_{(\alpha,u)}(a,0)h(u)$$

$$- \int_0^a R_{(\alpha,u)}(a,s)\phi\left(s,f(s,u(\beta(s))),\int_0^s g(s,\tau,u(\gamma(\tau)))d\tau\right)ds$$

$$- \sum_{i=1}^m R_{(\alpha,u)}(a,t_i)\{I_i(u(t_i)) - I_i(0) + I_i(0)\}]$$

$$[u_1 - R(\alpha,v)(a,0)u_0 + R(\alpha,v)(a,0)h(v)$$

$$- \int_0^a R(\alpha,v)(a,s)\phi(s,f(s,v(\beta(s))), \int_0^s g(s,\tau,v(\gamma(\tau)))d\tau)ds$$

$$- \sum_{i=1}^m R_{(\alpha,v)}(a,t_i)\{I_i(v(t_i)) - I_i(0) + I_i(0)]\}])\|_Y\} \max_{\tau \in J} \|u(\tau - v(\tau))\|$$

$$\leq 2a^2 KM_1M_2\{\|u_1\|_Y + M_0(\|u_0\|_Y + k_1 + k_7\theta + k_8 a + \xi)\}$$
$$\times \max_{\tau \; in J} \|u(\tau) - v(\tau)\|.$$

By using the Lemmas 10.1 and 10.2, $A_3 - A_6$ and A_8 we conclude

$$I_3 = \int_0^t \{\|R_{(\alpha,u)}(t,s)\phi(s,f(s,u(\beta(s))), \int_0^s g(s,\tau,u(\gamma(\tau)))d\tau)$$

$$- R_{(\alpha,v)}(t,s)\phi(s, f(s, u(\beta(s))), \int_0^s g(s, \tau, u(\gamma(\tau)))d\tau)\|$$

$$+ \|R_{(\alpha,v)}(t,s)\phi(s, f(s, u(\beta(s))), \int_0^s g(s, \tau, u(\gamma(\tau)))d\tau)$$

$$- R_{(\alpha,v)}(t,s)\phi(s, f(s, u(\beta(s))), \int_0^s g(s, \tau, u(\gamma(\tau)))d\tau)\|\}ds$$

$$\leq Ka(k_7\theta + k_8 \max_{\tau \in J})\|u(\tau) - v(\tau)\|$$

$$+ M_0 k_7 \int_0^t \{\|f(s, u(\beta(s)) - f(s, v(\beta(s)/c_p)))\|$$

$$+ \int_0^s g(s, \tau, u(\gamma(\tau))) - g(s, \tau, v(\gamma(\tau)))\|d\tau\}ds$$

$$\leq Ka(k_7\theta + k_8 \max_{\tau \in J})\|u(\tau) - v(\tau)\|$$

$$+ M_0 k_7 \int_0^t \{k_5 \sum_{p=1}^r \|u(\beta_p(s)) - v(\beta_p(s))\|(\beta_p'(s)/c_p)$$

$$+ k_3 \sum_{q=1}^k \|u(\gamma_q(s)) - v(\gamma_q(s))\|(\gamma_p'(s)/b_q)\}ds$$

$$\leq \{Ka(k_7\theta + k_8a) + M_0 k_7\rho\} \max_{\tau \in J}\|u(\tau) - v(\tau)\|.$$

Again from Lemma 10.1, A_7 and A_8, we obtain

$$I_4 \leq \sum_{i=1}^m \{\|R_{(\alpha,u)}(t,t_i)I_i(u(t_i)) - R_{(\alpha,v)}(t,t_i)I_i(u(t_i))\|$$

$$+ \|R_{(\alpha,v)}(t,t_i)I_i(u(t_i))$$

$$- R_{(\alpha,v)}(t,t_i)I_i(u(t_i))\|\}$$

$$\leq \left\{ K \sum_{i=1}^m (l_i\delta + \|I_i(0)\|)a + M_0 \sum_{i=1}^m l_i \right\}$$

$$+ \max_{\tau \in J}\|u(\tau) - v(\tau)\|.$$

As a result we conclude that

$$\|Pu_\mu(t) - Pv_\mu(t)\| \leq \sum_{j=1}^4 I_j \leq \sum_{j=1}^4 \lambda_j \max \|u(\tau) - v(\tau)\|.$$

Thus, P is a contraction mapping and hence there exists a unique fixed point $u \in X$, provided that $Pu(t) = u(t)$. Any fixed point of P represents a mild solution of (10.4) – (10.6) on J satisfying $u(a) = u_1$. Finally, we conclude that the system (10.4) – (10.6) is controllable on J. $\square$

Remark 10.2. [Debbouche and Baleanu (2011)] Let $u_a(u_0 - h(u); \mu)$ be the state value of (10.4) – (10.6) at terminal time a corresponding to the control μ and the nonlocal initial value $u_0 - h(u) \in P$ is an abstract phase space described axiomatically.

We introduce as well the set

$$\Re(a, u_0 - h(u)) = \{u_a(u_0 - h(u); \mu)(0) : \mu \in L_2(J, U)\},$$

which is called the reachable set of system (10.4) – (10.6) at terminal time a and its closure in X is denoted by $\overline{\Re(a, u_0 - h(0))}$.

We recall that the fractional nonlocal impulsive control system (10.4) – (10.6) is said to be approximately controllable on the interval J if $\overline{\Re(a, u_0 - h(0))} = X$, see [Subalakshmi and Balachandran (2009); Sakthivel and Anandhi (2009); Debbouche and Baleanu (2011)].

Chapter 11

Approximate Controllability of Sobolev Type Nonlocal Fractional Stochastic Dynamic Systems in Hilbert Spaces

The controllability of the deterministic and stochastic dynamical control systems in infinite dimensional spaces is an well-developed field. We recall that the theory of controllability for nonlinear fractional dynamical systems started to be developed intensively during the last few years (see for example [Sukavanam and Kumar (2011); Sukavanam and Tomar (2007); Kerboua et al. (2013)] and the references therein). The exact controllability for semilinear fractional order system, such that the nonlinear term is independent of the control function, was reported in [Bashirov and Mahmudov (1999); Debbouche and Baleanu (2011, 2012); Sakthivel et al. (2012a)]. In these works it was proved the exact controllability by assuming that the controllability operator has an induced inverse on a quotient space. If the semigroup associated with the system is compact, then the controllability operator is also compact,as a result the induced inverse does not exist mainly because the state space is infinite dimensional [Yan (2012a)]. Hence, the concept of exact controllability is too strong and implicitly has limited applicability. as it is known the approximate controllability is a weaker concept than the complete one and it is completely adequate in applications for these control systems.

In [Dauer and Mahmudov (2002); Triggiani (1977)] the approximate controllability of first order delay control systems was reported when nonlinear term is a function of both state function and control function by making the assumption that the corresponding linear system be approximately controllable. To prove the approximate controllability of a first order system, with or without delay, a relation between the reachable set of a semilinear system and that of the corresponding linear system is proved in [Bian (1998); Chukwu and Lenhart (1991); Jeong et al. (2007); Jeong and Kim (2009); Yan (2012b)]. For the readers we suggest several papers devoted to the approximate controllability for semilinear control systems, such that the nonlinear term is independent of control function [Kumar and Sukavanam (2012); Sakthivel and Ren (2012); Sakthivel et al. (2011); Sukavanam and Kumar (2011)]. Besides, there are many interesting results on the theory and applications of stochastic differential equations (see [Bashirov and Mahmudov (1999); Cao et al. (2011, 2012); Chang et al. (2011); Mao (1997); Sakthivel et al. (2013)] and the references therein).

Taking into account that the deterministic models often fluctuate due to noise, so are encouraged to move from deterministic control to stochastic control problems.

Some results on the approximate controllability of fractional stochastic systems can be seen in [Mahmudov (2003); Kerboua et al. (2013)], as well as on the existence and controllability results of fractional evolution equations of Sobolev type [Li et al. (2012)].

In [Sakthivel et al. (2012b)] the authors investigated the approximate controllability of a class of dynamic control systems described by nonlinear fractional stochastic differential equations in Hilbert spaces.

More details about some related resulted reported in this field can be found in [Sukavanam and Kumar (2011); Sukavanam and Tomar (2007); Debbouche et al. (2012); Feckan et al. (2013); Kerboua et al. (2013)] and the references therein.

11.0.1 *The Problem*

Below we discuss the nonlocal fractional stochastic system of Sobolev type [Kerboua et al. (2013)]

$$_{0}C_{t}^{q}[Lx(t)] = Mx(t) + Bu(t) + f(t, x(t)) + \sigma(t, x(t))\frac{dw(t)}{dt}, \qquad (11.1)$$

$$x(0) + g(x(t)) = x_{0}. \qquad (11.2)$$

Here $_{0}C_{t}^{q}$ means the Caputo fractional derivative of order q, $0 < q \leq 1$, and $t \in J = [0, b]$. Let X and Y be two Hilbert spaces and the state $x(\cdot)$ takes its values in X. We assume that L and M are defined on domains contained in X and ranges contained in Y, the control function $u(\cdot)$ belongs to the space $L_{\Gamma}^{2}(J, U)$, a Hilbert space of admissible control functions with U as a Hilbert space and B is a bounded linear operator from U into Y.

We suppose that $f : J \times X \to Y, g : C(J : X) \to Y$ and $\sigma : J \times X \to L_2^0$ are appropriate functions; x_0 is Γ_0 measurable X-valued random variables independent of w. Γ, Γ_0, L_2^0 and w will be specified later [Kerboua et al. (2013)].

Next, we make some assumptions on the operators L and M [Kerboua et al. (2013)].

(A_1) L and M are linear operators, and M is closed.
(A_2) $D(L) \subset D(M)$ and L is bijective.
(A_3) $L^{-1} : Y \to D(L) \subset X$ is a linear compact operator.

From (A_3), we conclude that L^{-1} is bounded operators and we denote by $C = \|L^{-1}\|$. Note (A_3) also implies that L is closed because L^{-1} is closed and injective, then its inverse is also closed [Kerboua et al. (2013)]. By using (A_1)–(A_3) and the closed graph theorem, we obtain the boundedness of the linear operator $ML^{-1} : Y \to Y$. Thus, ML^{-1} generates a semigroup $\{S(t) := e^{ML^{-1}t}, t \geq 0\}$. We also assume that $M_0 := \sup_{t \geq 0} \|S(t)\| < \infty$. As a result, we rewrite the problem 11.1–11.2 as an equivalent integral equation, namely [Kerboua et al. (2013)]

$$Lx(t) = L[x_0 - g(x)] + \frac{1}{\Gamma(q)} \int_0^t (t - s)^{q-1} [Mx(s) + Bu(s) + f(s, x(s))] ds$$

$$+ \frac{1}{\Gamma(q)} \int_0^t (t - s)^{q-1} \sigma(s, x(s)) dw(s), \tag{11.3}$$

assuming that the integral in (11.3) exists. Before introducing the definition of mild solution of (11.1)–(11.2), we present the following definitions, corollaries, lemmas and notations.

Let (Ω, Γ, P) be a complete probability space equipped with a normal filtration $\Gamma_t, t \in J$ satisfying the usual conditions (right continuous and Γ_0 containing all P-null sets). We consider four real separable spaces X, Y, E and U, and Q-Wiener process on (Ω, Γ_b, P) with the linear bounded covariance operator Q fulfilling $trQ < \infty$. We suppose that there exists a complete orthonormal system $\{e_n\}_{n \geq 1}$ in E, a bounded sequence of non-negative real numbers $\{\lambda_n\}$ such that $Qe_n = \lambda_n e_n, n = 1, 2, ...$, and a sequence $\{\beta_n\}_{n \geq 1}$ of independent Brownian motions fulfilling [Kerboua et al. (2013)]

$$\langle w(t), e \rangle = \sum_{n=1}^{\infty} \sqrt{\lambda_n} \langle e_n, e \rangle \beta_n(t), \quad e \in E, \ t \in J$$

and $\Gamma_t = \Gamma_t^w$, where Γ_t^w is the sigma algebra generated by $\{w(s) : 0 \leq s \leq t\}$. Let $L_2^0 = L_2(Q^{1/2}E; X)$ be the space of all Hilbert–Schmidt operators from $Q^{1/2} E$ to X equipped with the inner product $\langle \psi, \pi \rangle L_2^0 = tr[\psi Q \pi^*]$. Now, let $L^2(\Gamma_b, X)$ be the Banach space of all Γ_b-measurable square integrable random variables with values in the Hilbert space X. [Kerboua et al. (2013)] Also, let $E(\cdot)$ denotes the expectation with respect to the measure P. Let $C(J; L^2(\Gamma, X))$ be the Hilbert space of continuous maps from J into $L^2(\Gamma, X)$ such that $\sup_{t \in J} E \|x(t)\|^2 < \infty$. Let $H_2(J; X)$ is a closed subspace of $C(J; L^2(\Gamma, X))$ consisting of measurable and Γ_t-adapted X-valued process $x \in C(J; L^2(\Gamma, X))$ endowed with the norm $\|x\|_{H_2} = (\sup_{t \in J} E \|x(t)\|_X^2)^{1/2}$. For more details, we guide the reader to ([Sakthivel et al. (2012b); Kerboua et al. (2013)] and the references therein).

The following results are utilized in our proofs [Kerboua et al. (2013)].

Lemma 11.1. ([Mahmudov (2003)]) *Let* $G : J \times \Omega \to L_2^0$ *be a strongly measurable mapping such that* $\int_0^b E \|G(t)\|_{L_2^0}^p dt < \infty$. *Then*

$$E \left\| \int_0^t G(s) dw(s) \right\|^p \leq L_G \int_0^t E \|G(s)\|_{L_2^0}^p ds$$

for all $0 \leq t \leq b$ *and* $p \geq 2$, *where* L_G *denotes the constant involving* p *and* b.

The next step is to present the mild solution of the problem 11.1–11.2 [Kerboua et al. (2013)].

Definition 11.1. (Compare with [Debbouche and El-Borai (2009); El-Borai (2002)] and [Feckan et al. (2013); Zhou and Jiao (2010b)])
A stochastic process $x \in H_2(J, X)$ is a mild solution of 11.1–11.2 if for each control $u \in L_\Gamma^2(J, U)$, it satisfies the following integral equation,

$$x(t) = \mathcal{S}(t) L[x_0 - g(x)] + \int_0^t (t-s)^{q-1} \mathcal{T}(t-s)[Bu(s) + f(s, x(s))] ds$$

$$+ \int_0^t (t-s)^{q-1} \mathcal{T}(t-s) \sigma(s, x(s)) dw(s), \tag{11.4}$$

where $\mathcal{S}(t)$ and $\mathcal{T}(t)$ are characteristic operators given by

$$\mathcal{S}(t) = \int_0^\infty L^{-1} \xi_q(\theta) S(t^q \theta) d\theta \text{ and } \mathcal{T}(t) = q \int_0^\infty L^{-1} \theta \xi_q(\theta) S(t^q \theta) d\theta.$$

Here, $S(t)$ is a C_0-semigroup generated by the linear operator $ML^{-1} : Y \to Y$; ξ_q is a probability density function defined on $(0, \infty)$, that is $\xi_q(\theta) \geq 0$, $\theta \in (0, \infty)$ and $\int_0^\infty \xi_q(\theta) d\theta = 1$.

Lemma 11.2. ([Yan (2012b,a); Zhou and Jiao (2010b)]) *The operators* $\{\mathcal{S}(t)\}_{t\geq 0}$ *and* $\{\mathcal{T}(t)\}_{t\geq 0}$ *are strongly continuous, i.e., for* $x \in X$ *and* $0 \leq t_1 < t_2 \leq b$, *we have* $||\mathcal{S}(t_2)x - \mathcal{S}(t_1)x|| \to 0$ *and* $||\mathcal{T}(t_2)x - \mathcal{T}(t_1)x|| \to 0$ *as* $t_2 \to t_1$.

Below we impose the following conditions on data of our problem:

(i) For any fixed $t \geq 0, \mathcal{T}(t)$ and $\mathcal{S}(t)$ are bounded linear operators, i.e., for any $x \in X$,

$$||\mathcal{T}(t)x|| \leq CM_0||x||, \quad ||\mathcal{S}(t)x|| \leq \frac{CM_0}{\Gamma(q)}||x||.$$

(ii) The functions $f : J \times X \to Y, \sigma : J \times X \to L_2^0$ and $g : C(J : X) \to Y$ satisfy linear growth and Lipschitz conditions. Moreover, there exist positive constants $N_1, N_2 > 0, L_1, L_2 > 0$ and $k_1, k_2 > 0$ such that

$$||f(t, x) - f(t, y)||^2 \leq N_1||x - y||^2, \quad ||f(t, x)||^2 \leq N_2(1 + ||x||^2),$$

$$||\sigma(t, x) - \sigma(t, y)||_{L_2^0}^2 \leq L_1||x - y||^2, \quad ||\sigma(t, x)||_{L_2^0}^2 \leq L_2(1 + ||x||^2),$$

$$||g(x) - g(y)||^2 \leq k_1||x - y||^2, \quad ||g(x)||^2 \leq k_2(1 + ||x||^2).$$

(iii) The linear stochastic system is approximately controllable on J.

For each $0 \leq t < b$, the operator $\alpha(\alpha I + \Psi_0^b)^{-1} \to 0$ in the strong operator topology as $\alpha \to 0^+$, where $\Psi_0^b = \int_0^b (b - s)^{2(q-1)} \mathcal{T}(b - s)BB^*\mathcal{T}^*(b - s)ds$ is the controllability Gramian, B^* is the adjoint of B and $\mathcal{T}^*(t)$ represents the adjoint of $\mathcal{T}(t)$.

We notice that the Sobolev type nonlocal linear fractional deterministic control system

$$_0C_t^q[Lx(t)] = Mx(t) + Bu(t), \ t \in J, \tag{11.5}$$

$$x(0) + g(x(t)) = x_0, \tag{11.6}$$

associated to 11.1–11.2 is approximately controllable on J iff the operator $\alpha(\alpha I + \Psi_0^b)^{-1} \to 0$ strongly as $\alpha \to 0^+$.

Definition 11.2. System 11.1–11.2 is approximately controllable on J if $\overline{\mathfrak{R}(b)} = L^2(\Omega, \Gamma_b, X)$, where

$$\mathfrak{R}(b) = \{x(b) = x(b, u) : u \in L_\Gamma^2(J, U)\},$$

where $L_\Gamma^2(J, U)$, denotes the closed subspace of $L_\Gamma^2(J \times \Omega; U)$, consisting of all Γ_t adapted, U-valued stochastic processes.

Lemma 11.3. [Kerboua et al. (2013)] *For any $\widetilde{x}_b \in L^2(\Gamma_b, X)$, there exists $\widetilde{\varphi} \in L^2_\Gamma(\Omega; L^2(0, b; L^0_2))$ such that $\widetilde{x}_b = E\widetilde{x}_b + \int_0^b \widetilde{\varphi}(s)dw(s)$.*

Now for any $\alpha > 0$ and $\widetilde{x}_b \in L^2(\Gamma_b, X)$, we define the control function in the following form

$$u^\alpha(t, x) = B^*(b - t)^{q-1}T^*(b - t)$$

$$\left(\left[(\alpha I + \Psi_0^b)^{-1}\{E\widetilde{x}_b - \mathcal{S}(b)L[x_0 - g(x)]\} + \int_0^t (\alpha I + \Psi_0^b)^{-1}\widetilde{\varphi}(s)dw(s) \right] \right.$$

$$-B^*(b - t)^{q-1}T^*(b - t)\int_0^t (\alpha I + \Psi_0^b)^{-1}(b - s)^{q-1}T(b - s)f(s, x(s))ds$$

$$-B^*(b - t)^{q-1}T^*(b - t)\int_0^t (\alpha I + \Psi_0^b)^{-1}(b - s)^{q-1}T(b - s)\sigma(s, x(s))dw(s).$$

Lemma 11.4. [Kerboua et al. (2013)] *There exist positive real constants $\hat{M}, \hat{N}$ such that for all $x, y \in H_2$, we have*

$$E\left\| u^\alpha(t, x) - u^\alpha(t, y) \right\|^2 \le \hat{M}E\left\| x(t) - y(t) \right\|^2, \qquad (11.7)$$

$$E\left\| u^\alpha(t, x) \right\|^2 \le \hat{N}\left(\frac{1}{b} + E\left\| x(t) \right\|^2 \right). \qquad (11.8)$$

Proof. *[Kerboua et al. (2013)]* We begin to prove (11.7). Let $x, y \in H_2$, from the Hölder's inequality, Lemma 11.1 and the assumption on the data, we obtain

$$E\left\| u^\alpha(t, x) - u^\alpha(t, y) \right\|^2$$

$$\le 3E\left\| B^*(b - t)^{q-1}T^*(b - t)(\alpha I + \Psi_0^b)^{-1}\mathcal{S}(b)L[g(x(t)) - g(y(t))] \right\|^2$$

$$+3E\left\| B^*(b - t)^{q-1}T^*(b - t)\int_0^t \hat{M}_1[f(s, x(s)) - f(s, y(s))]ds \right\|^2$$

$$+3E\left\| B^*(b - t)^{q-1}T^*(b - t)\int_0^t \hat{M}_1[\sigma(s, x(s)) - \sigma(s, y(s))]dw(s) \right\|^2$$

$$\le \frac{3}{\alpha^2}\left\| B \right\|^2 (b)^{2q-2}(CM_0)^2\left(\frac{CM_0}{\Gamma(q)} \right)^2\left\| L \right\|^2 k_1 E\left\| x(t) - y(t) \right\|^2$$

$$+\frac{3}{\alpha^2}\left\| B \right\|^2 (b)^{2q-2}\left(\frac{CM_0}{\Gamma(q)} \right)^4 \frac{b^{2q-1}}{(2q - 1)}N_1\int_0^t E\left\| x(s) - y(s) \right\|^2 ds$$

$$+\frac{3}{\alpha^2}\left\| B \right\|^2 (b)^{2q-2}\left(\frac{CM_0}{\Gamma(q)} \right)^4 \frac{b^{2q-1}}{(2q - 1)}L_1\int_0^t E\left\| x(s) - y(s) \right\|^2 ds$$

$$\le \hat{M}E\left\| x(t) - y(t) \right\|^2,$$

where
$$\hat{M}_1 = (\alpha I + \Psi_0^b)^{-1}(b-s)^{q-1}\mathcal{T}(b-s)$$

and
$$\hat{M} = \frac{3}{\alpha^2}\|B\|^2(b)^{2q-2}\left\{(CM_0)^2\left(\frac{CM_0}{\Gamma(q)}\right)^2\|L\|^2 k_1 \right.$$
$$\left. + \left(\frac{CM_0}{\Gamma(q)}\right)^4 \frac{b^{2q-1}}{(2q-1)}b[N_1 + L_1]\right\}.$$

The proof of the inequality (11.7) can be established in a similar way to that of (11.6).

11.0.2 *Approximate Controllability*

For any $\alpha > 0$, let us define the operator $F_\alpha : H_2 \to H_2$ as

$$F_\alpha x(t) = \mathcal{S}(t)L[x_0 - g(x)] + \int_0^t (t-s)^{q-1}\mathcal{T}(t-s)[Bu^\alpha(s,x) + f(s,x(s))]ds$$
$$+ \int_0^t (t-s)^{q-1}\mathcal{T}(t-s)\sigma(s,x(s))dw(s). \tag{11.9}$$

Lemma 11.5. [Kerboua et al. (2013)] *For any $x \in H_2$, $F_\alpha(x)(t)$ is continuous on J in L^2-sense.*

Proof. [*Kerboua et al.(2013)*] Let $0 \le t_1 < t_2 \le b$. Then for any fixed $x \in H_2$, from (11.9), we have

$$E\|(F_\alpha x)(t_2) - (F_\alpha x)(t_1)\|^2 \le 4\left[\sum_{i=1}^4 E\|\Pi_i^x(t_2) - \Pi_i^x(t_1)\|^2\right].$$

We begin with the first term, we get
$$E\|\Pi_1^x(t_2) - \Pi_1^x(t_1)\|^2 = E\|(\mathcal{S}(t_2) - \mathcal{S}(t_1))L[x_0 - g(x)]\|^2$$
$$\le \|L\|^2[\|x_0\|^2 + k_2(1 + \|x\|^2)]E\|\mathcal{S}(t_2) - \mathcal{S}(t_1)\|^2.$$

The strong continuity of $\mathcal{S}(t)$ implies that the right hand side of the last inequality tends to zero as $t_2 - t_1 \to 0$.

Next, it follows from Hölder's inequality and assumptions on the data that
$$E\|\Pi_2^x(t_2) - \Pi_2^x(t_1)\|^2$$
$$\le E\left\|\int_0^{t_1} (t_1 - s)^{q-1}(\mathcal{T}(t_2 - s) - \mathcal{T}(t_1 - s))Bu^\alpha(s,x)ds\right\|^2$$

$$+ E \left\| \int_0^{t_1} ((t_2 - s)^{q-1} - (t_1 - s)^{q-1}) T(t_2 - s) B u^\alpha(s, x) ds \right\|^2$$

$$+ E \left\| \int_{t_1}^{t_2} (t_2 - s)^{q-1} T(t_2 - s) B u^\alpha(s, x) ds \right\|^2$$

$$\leq \frac{t_1^{2q-1}}{2q - 1} \int_0^{t_1} E \left\| (T(t_2 - s) - T(t_1 - s)) B u^\alpha(s, x) ds \right\|^2$$

$$+ \left(\frac{C M_0 \|B\|}{\Gamma(q)} \right)^2 \left(\int_0^{t_1} ((t_2 - s)^{q-1} - (t_1 - s)^{q-1})^2 ds \right)$$

$$\times \left(\int_0^{t_1} E \left\| u^\alpha(s, x) \right\|^2 ds \right) + \frac{(t_2 - t_1)^{2q-1}}{1 - 2q} \int_{t_1}^{t_2} E \left\| u^\alpha(s, x) \right\|^2 ds.$$

Also, we have

$$E \left\| \Pi_3^x(t_2) - \Pi_3^x(t_1) \right\|^2$$

$$\leq E \left\| \int_0^{t_1} (t_1 - s)^{q-1} (T(t_2 - s) - T(t_1 - s)) f(s, x(s)) ds \right\|^2$$

$$+ E \left\| \int_0^{t_1} ((t_2 - s)^{q-1} - (t_1 - s)^{q-1}) T(t_2 - s) f(s, x(s)) ds \right\|^2$$

$$+ E \left\| \int_{t_1}^{t_2} (t_2 - s)^{q-1} T(t_2 - s) f(s, x(s)) ds \right\|^2$$

$$\leq \frac{t_1^{2q-1}}{2q - 1} \int_0^{t_1} E \left\| (T(t_2 - s) - T(t_1 - s)) f(s, x(s)) ds \right\|^2$$

$$+ \left(\frac{C M_0}{\Gamma(q)} \right)^2 \left(\int_0^{t_1} ((t_2 - s)^{q-1} - (t_1 - s)^{q-1})^2 ds \right)$$

$$\times \left(\int_0^{t_1} E \left\| f(s, x(s)) \right\|^2 ds \right) + \frac{(t_2 - t_1)^{2q-1}}{1 - 2q} \left(\frac{C M_0}{\Gamma(q)} \right)^2$$

$$\times \int_{t_1}^{t_2} E \left\| f(s, x(s)) \right\|^2 ds.$$

Furthermore, we utilize the Lemma 11.1 and the previous assumptions and we obtain

$$E \left\| \Pi_4^x(t_2) - \Pi_4^x(t_1) \right\|^2$$

$$\leq E \left\| \int_0^{t_1} (t_1 - s)^{q-1} (T(t_2 - s) - T(t_1 - s)) \sigma(s, x(s)) dw(s) \right\|^2$$

$$+ E \left\| \int_0^{t_1} ((t_2 - s)^{q-1} - (t_1 - s)^{q-1}) T(t_2 - s) \sigma(s, x(s)) dw(s) \right\|^2$$

$$+ E \left\| \int_{t_1}^{t_2} (t_2 - s)^{q-1} T(t_2 - s) \sigma(s, x(s)) dw(s) \right\|^2$$

$$\leq L_\sigma \frac{t_1^{2q-1}}{2q-1} \int_0^{t_1} E \left\| (T(t_2 - s) - T(t_1 - s)) \sigma(s, x(s)) ds \right\|^2$$

$$+ L_\sigma \left(\int_0^{t_1} ((t_2 - s)^{q-1} - (t_1 - s)^{q-1})^2 ds \right)$$

$$\times \left(\int_0^{t_1} E \left\| T(t_2 - s) \sigma(s, x(s)) \right\|^2 ds \right)$$

$$+ L_\sigma \frac{(t_2 - t_1)^{2q-1}}{1 - 2q} \left(\frac{CM_0}{\Gamma(q)} \right)^2 \int_{t_1}^{t_2} E \left\| T(t_2 - s) \sigma(s, x(s)) \right\|^2 ds.$$

By using the strong continuity of $T(t)$ and Lebesgue's dominated convergence theorem, we conclude that the right-hand side of the above inequalities tends to zero as $t_2 - t_1 \to 0$. Thus, it is clear that $F_\alpha(x)(t)$ is continuous from the right of $[0, b)$. A similar type of proof shows that it is continuous from the left of $(0, b]$.

Theorem 11.1. [Kerboua et al. (2013)] *Assume hypotheses (i) and (ii) are satisfied. Then, the system 11.1–11.2 has a mild solution on J.*

Proof. [*Kerboua et al.(2013)*] We prove the existence of a fixed point of the operator F_α by using the contraction mapping principle. First, we show that $F_\alpha(H_2) \subset H_2$. Let $x \in H_2$. From (11.9), we obtain

$$E \left\| F_\alpha x(t) \right\|^2 \leq 4 \left[\sup_{t \in J} \sum_{i=1}^{4} E \left\| \Pi_i^x(t) \right\|^2 \right]. \tag{11.10}$$

Using the assumptions (i)–(ii), the Lemma 11.4, by standard computations we obtain

$$\sup_{t \in J} E \left\| \Pi_1^x(t) \right\|^2 \leq C^2 M_0^2 \| L \|^2 [\| x_0 \|^2 + k_2 (1 + \| x \|^2)] \tag{11.11}$$

and

$$\sup_{t \in J} \sum_{i=2}^{4} E \left\| \Pi_i^x(t) \right\|^2 \leq \left(\frac{CM_0}{\Gamma(q)} \right)^2 \frac{b^{2q-1}}{2q-1} \| B \|^2 \hat{N} \left(\frac{1}{b} + \| x \|_{H_2}^2 \right)$$

$$+ \left(\frac{CM_0}{\Gamma(q)} \right)^2 \left[\frac{b^{2q-1}}{2q-1} N_2 - \frac{b^{2q-1}}{2q-1} L_2 L_\sigma \right] (1 + \| x \|_{H_2}^2). \tag{11.12}$$

Hence (11.10)–(11.12) imply that $E \| F_\alpha x \|_{H_2}^2 < \infty$. By Lemma 3.1, $F_\alpha x \in H_2$. Thus for each $\alpha > 0$, the operator F_α maps H_2 into itself. Next, we use the Banach fixed point theorem to prove that F_α has a unique fixed point in H_2. We claim that there exists a natural n such that F_α^n is a contraction on H_2. Indeed, let $x, y \in H_2$, we have

$$E \left\| (F_\alpha x)(t) - (F_\alpha y)(t) \right\|^2 \leq 4 \sum_{i=1}^{4} E \left\| \Pi_i^x(t) - \Pi_i^y(t) \right\|^2$$

$$\leq 4k_1 C^2 M_0^2 \|L\|^2 E \|x(t) - y(t)\|^2$$

$$+ 4 \left(\frac{C M_0}{\Gamma(q)} \right)^2 \left[\hat{M} \|B\|^2 \frac{b^{2q-1}}{2q-1} + \frac{b^{2q-1}}{2q-1} N_1 + \frac{b^{2q-1}}{2q-1} L_1 L_\sigma \right] E \|x(t) - y(t)\|^2.$$

Hence, we obtain a positive real constant $\gamma(\alpha)$ such that

$$E \left\| (F_\alpha x)(t) - (F_\alpha y)(t) \right\|^2 \leq \gamma(\alpha) E \|x(t) - y(t)\|^2, \tag{11.13}$$

for all $t \in J$ and all $x, y \in H_2$. For any natural number n, it follows from successive iteration of above inequality (11.13) that, by taking the supremum over J,

$$\left\| (F_\alpha^n x)(t) - (F_\alpha^n y)(t) \right\|_{H_2}^2 \leq \frac{\gamma^n(\alpha)}{n!} \|x - y\|_{H_2}^2. \tag{11.14}$$

For any fixed $\alpha > 0$, for sufficiently large n, $\frac{\gamma^n(\alpha)}{n!} < 1$. It follows from (11.14) that F_α^n is a contraction mapping, so that the contraction principle ensures that the operator F_α has a unique fixed point x_α in H_2, which is a mild solution of 11.1–11.2.

Theorem 11.2. [Kerboua et al. (2013)] *Suppose that the assumptions (i)–(iii) hold. Further, if the functions f and σ are uniformly bounded and $\{\mathcal{T}(t) : t \geq 0\}$ is compact, then the system 11.1–11.2 is approximately controllable on J.*

Proof. [*Kerboua et al.(2013)*] Let x_α be a fixed point of F_α . By using the stochastic Fubini theorem, it can be easily seen that

$$x_\alpha(b) = \widetilde{x}_b - \alpha(\alpha I + \Psi)^{-1}(E\widetilde{x}_b - \mathcal{S}(b)L[x_0 - g(x)])$$

$$+ \alpha \int_0^b (\alpha I + \Psi_s^b)^{-1}(b - s)^{q-1} T(b - s) f(s, x_\alpha(s)) ds$$

$$+ \alpha \int_0^b (\alpha I + \Psi_s^b)^{-1}[(b - s)^{q-1} T(b - s) \sigma(s, x_\alpha(s)) - \widetilde{\varphi}(s)] dw(s).$$

It follows from the assumption on f, g and σ that there exists $\hat{D} > 0$ such that

$$\|f(s, x_\alpha(s))\|^2 + \|g(x_\alpha(s))\|^2 + \|\sigma(s, x_\alpha(s))\|^2 \leq \hat{D} \tag{11.15}$$

for all $s \in J$. Then, there is a subsequence still denoted by $\{f(s, x_\alpha(s)),$ $g(x_\alpha(s)), \sigma(s, x_\alpha(s))\}$ which converges weakly to some $\{f(s), g(s), \sigma(s)\}$ in $Y^2 \times L_2^0$.

From the above equation, we have

$$
\begin{aligned}
E \left\| x_\alpha(b) - \tilde{x}_b \right\|^2 \\
\leq 8E \left\| \alpha(\alpha I + \Psi_0^b)^{-1}(E\tilde{x}_b - \mathcal{S}(b)Lx_0) \right\|^2 \\
+ 8E \| \alpha(\alpha I + \Psi_0^b)^{-1} \|^2 \| \mathcal{S}(b)L(g(x_\alpha(s)) - g(s)) \|^2 \\
+ 8E \| \alpha(\alpha I + \Psi_0^b)^{-1} \|^2 \| \mathcal{S}(b)Lg(s) \|^2 \\
+ 8E \left(\int_0^b (b-s)^{q-1} \left\| \alpha(\alpha I + \Psi_s^b)^{-1} \widetilde{\varphi}(s) \right\|_{L_2^0}^2 ds \right) \\
+ 8E \left(\int_0^b (b-s)^{q-1} \left\| \alpha(\alpha I + \Psi_s^b)^{-1} \right\| \| \mathcal{T}(b-s)(f(s, x_\alpha(s)) - f(s)) \| ds \right)^2 \\
+ 8E \left(\int_0^b (b-s)^{q-1} \left\| \alpha(\alpha I + \Psi_s^b)^{-1} \mathcal{T}(b-s) f(s) \right\| ds \right)^2 \\
+ 8E \left(\int_0^b (b-s)^{q-1} \left\| \alpha(\alpha I + \Psi_s^b)^{-1} \right\| \| \mathcal{T}(b-s)(\sigma(s, x_\alpha(s)) - \sigma(s)) \|_{L_2^0}^2 ds \right) \\
+ 8E \int_0^b (b-s)^{q-1} \left\| \alpha(\alpha I + \Psi_s^b)^{-1} \mathcal{T}(b-s) \sigma(s) \right\|_{L_2^0}^2 ds.
\end{aligned}
$$

By using the assumption (iii), for all $0 \leq s < b$ the operator $\alpha(\alpha I + \Psi_s^b)^{-1} \to 0$ strongly as $\alpha \to 0^+$ and $\|\alpha(\alpha I + \Psi_s^b)^{-1}\| \leq 1$. By using the Lebesgue dominated convergence theorem and the compactness of both $\mathcal{S}(t)$ and $\mathcal{T}(t)$ we obtain $E\|x_\alpha(b) - \tilde{x}_b\|^2 \to 0$ as $\alpha \to 0^+$. Hence, the approximate controllability of 11.1–11.2 was proved.

Bibliography

Abada, N., Benchohra, N.M. and Hammouche, H. (2009). Existence and controllability results for nondensely defined impulsive semilinear functional differential inclusions, *J. Differential Eqs.* **246**, pp. 3834–3863.

Abdeljawad, T. (2011). On Riemann and Caputo fractional differences, *Comp. Math. Appl.* **62**, pp.1602–1611.

Abdeljawad, T. and Baleanu, D. (2011a). Fractional differences and integration by parts, *J. Comput. Anal. Appl.*, **13**, pp. 574–582.

Abdeljawad, T. and Baleanu, D. (2011b). Caputo q-fractional initial value problems and a q-analogue Mittag-Leffler function, *Commun. Nonlin. Sci.* **(16)**, pp. 4682-4688.

Abdeljawad, T., Baleanu, D., Jarad, F., Mustafa, O.G. and Trujillo, J.J. (2010). A Fite type result for sequential fractional differential equations, *Dyn. Syst. Appl.* **19**, pp. 383–394.

Abdeljawad, T. Baleanu, D., Jarad, F. and Agarwal, R.P. (2013). Fractional sums and differences with binomial coefficients, *Discrete Dynam. Nature Soc.*, Article Number: 104173, DOI:10.1155/2013/104173.

Ablowitz, M.J. and Fokas, A.S. (2003). Complex variables. Introduction and applications, Cambridge Univ. Press, Cambridge.

Agarwal, R.P. (2000). Difference equations and inequalities, Theory, Methods, and Applications. Marcel Dekker, New York.

Agarwal, R.P., Benchohra, M. and Hamani, S. (2010a). A survey on existence results for boundary value problems of nonlinear fractional differential equations and inclusions, *Acta Appl. Math.* **109**, pp. 973–1033.

Agarwal, R.P., Djebali, S., Moussaoui T. and Mustafa, O.G. (2007a). On the asymptotic integration of nonlinear differential equations, *J. Comput. Appl. Math.* **202**, pp. 352–376.

Agarwal, R.P., Djebali, S., Moussaoui T., Mustafa, O.G. and Rogovchenko, Y.V. (2007b). On the asymptotic behavior of solutions to nonlinear ordinary differential equations, *Asympt. Anal.* **54**, pp. 1–50.

Agarwal, R.P. and Lakshmikantham, V. (1993). Uniqueness and nonuniqueness criteria for ordinary differential equations. World Scientific, New Jersey.

Agarwal, R.P., Lakshmikantham, V. and Nieto, J.J. (2010b). On the concept of solution for fractional differential equations with uncertainty, *Nonlinear Anal.* **72**, pp. 2859–2862.

Agarwal, R.P. and Mustafa, O.G. (2007). A Riccatian approach to the decay of solutions of certain semi-linear PDE's, *Appl. Math. Lett.* **20**, pp. 1206–1210.

Agarwal, R.P., Zhou, Y. and He, Y. (2010c). Existence of fractional neutral functional differential equations, *Comp. Math. Appl.* **59**, pp. 1095–1100.

Agrawal, O.P. (2002). Formulation of Euler-Lagrange equations for fractional variational problems. *J. Math. Anal. Appl.* **272**, pp. 369–379.

Ahmad, B., and Nieto, J.J. (2012). Anti-periodic fractional boundary value problems with nonlinear term depending on lower order derivative, *Frac. Cal. Appl. Anal.* **15**, pp. 451–462.

Ahmad, B., Nieto, J.J., Alsaedi, A. and Mohamad, N. (2013). On a new class of antiperiodic fractional boundary value problems, *Abstr. Appl. Anal.* Article Number: 606454.

Anastassiou, G.A. (2011). Advances on Fractional Inequalities. Springer, New York.

Atici, F.M. and Eloe, P.W. (2009a). Initial value problems in discrete fractional calculus, *Proc. Amer. Math. Soc.* **137**, pp. 981–989.

Atici, F.M. and Eloe, P.W. (2009b). Discrete fractional calculus with the nabla operator, *Electronic J. Qual. Th. Diff. Eqs.* **2009**, pp. 1–12.

Atici, F.M. and Eloe, P.W. (2007). A transform method in discrete fractional calculus, *Inter. J. Diff. Eqs.* **2**, pp. 165–176.

Ahlfors, L.V. (1979). Complex Analysis, Third Edition, McGraw-Hill, New York.

Araya, D. and Lizama, C. (2008). Almost automorphic mild solutions to fractional differential equations, *Nonlinear Anal.* **69**, pp. 3692–3705.

Athanassov, Z.S. (1990). Uniqueness and convergence of successive approximations for ordinary differential equations, *Math. Japon.* **35**, pp. 351–367.

Avramescu, C. (1969). Sur l'existence des solutions convergentes de systèmes d'équations différentielles non linéaires, *Ann. Mat. Pura Appl.* **81**, pp. 147–168.

Babakhani, A. and Baleanu, D. (2011). Existence of positive solutions for a class of delay fractional differential equations with generalization to N-term, *Abstr. Appl. Anal.* **2011**, Article ID 391971, http://dx.doi.org/10.1155/2011/391971.

Babakhani, A. and Daftardar-Gejji, V. (2003). Existence of positive solutions for multi-term non-autonomous fractional differential equations, *J. Math. Anal. Appl.* **278**, pp. 434–442.

Babakhani, A. and Daftardar-Gejji, V. (2006). Existence of positive solutions for multi-term non-autonomous fractional differential equations with polynomial coefficients, *Electronic J. Diff. Eqs.* **129**, pp. 1–12.

Babakhani, A. and Enteghami, E. (2009). Existence of positive solutions for multiterm fractional differential equations of finite delay with polynomial coefficients, *Abstract Appl. Anal.* **(2009)**, Article ID 768920.

Balachandran, K. and Park, J.Y. (2009). Controllability of fractional integrodifferential systems in Banach spaces, *Nonlinear Anal., Hybrid Systems* **3**, pp. 363–367.

Balachandran, K. Kiruthika, S. and Trujillo, J.J. (2011). On fractional impulsive equations of Sobolev type with nonlocal condition in Banach Spaces, *Comp. Math. Appl.* **62**, pp. 1157-1165.

Balachandran, K. Kiruthika, S. and Trujillo, J.J. (2012). Relative controllability of fractional dynamical systems with multiple delays in control, *Comp. Math. Appl.* **64**, pp. 3037-3045.

Balachandran, K. and Sakthivel, R. (2001). Controllability of integrodifferential systems in Banach spaces, *Appl. Math. Comput.* **118**, pp. 63–71.

Baleanu, D. and Avkar, T. (2004). Lagrangians with linear velocities within Riemann-Liouville fractional derivatives, *Nuovo Cimento B* **119**, pp. 73–79.

Baleanu, D. and Muslih, S.I. (2006). Lagrangian formulation of classical fields within Riemann-Liouville fractional derivatives, *Physica Scripta* **72**, pp. 119–121.

Baleanu, D., Jarad, F. and Muslih, S. (2006). Difference discrete and fractional variational principles, *Facta Univ.: Phys. Chem. Tech.* **4**, pp. 175–183.

Baleanu, D. and Agrawal, O.P. (2006). Fractional Hamilton formalism within Caputo's derivative, *Czech. J. Phys.* **56**, pp. 1087–1092.

Baleanu, D. (2009). Fractional variational principles in action, *Phys. Scr.* **T136**, pp. 014006.

Baleanu, D. and Golmankhaneh, A.K. (2010). On electromagnetic field in fractional space, *Nonlinear Anal., RWA.* **11**, pp. 288–292.

Baleanu, D. and Mustafa, O.G. (2009). On the asymptotic integration of a class of sublinear fractional differential equations, *J. Math. Phys.* **50**, pp. 123520.

Baleanu, D. and Mustafa, O.G. (2010a). On the global existence of solutions to a class of fractional differential equations, *Comp. Math. Appl.* **59**, pp. 1835–1841.

Baleanu, D., Mustafa, O.G. and Agarwal, R.P. (2010b). On the solution set for a class of sequential fractional differential equation, *J. Phys. A* **43**, pp. 385209.

Baleanu, D., Mustafa, O.G. and Agarwal, R.P. (2010c). Asymptotically linear solutions of some linear fractional differential equations, *Abstract Appl. Anal.* **2010**, Article ID 865139, http://dx.doi.org/10.1155/2010/865139.

Baleanu, D., Mustafa, O.G. and Agarwal, R.P. (2010d). An existence result for a superlinear fractional differential equation, *Appl. Math. Lett.* **23**, pp. 1129–1132.

Baleanu, D., Sadati, S.J., Ghaderi, R., Ranjbar, N.A., Abdeljawad (Maraaba), T. and Jarad, F. (2010e). Razumikhin stability theorem for fractional systems with delay, *Abstract Appl. Anal.* **(2010)**, Article ID 124812.

Baleanu, D. and Mustafa, O.G. (2011). On the existence interval for the initial value problem of a fractional differential equation, *Hacettepe J. Math. Stat.* **40**, pp. 581–587.

Baleanu, D., Agarwal, R.P., Mustafa, O.G. and Coşulschi, M. (2011a). Asymptotic integration of some nonlinear differential equations with fractional time derivative, *J. Phys. A* **44**, pp. 055203.

Baleanu, D., Mustafa, O.G. and Agarwal, R.P. (2011b). Asymptotic integration of $(1 + \alpha)$-order fractional differential equations, *Comp. Math. Appl.* **62**, pp. 1492–1500.

Baleanu, D., Mustafa, O.G. and Agarwal, R.P. (2011c). On L^p-solutions for a class of sequential fractional differential equations, *Appl. Math. Comput.* **218**, pp. 2074–2081.

Baleanu, D., Mustafa, O.G. and O'Regan, D. (2011d). A Nagumo-like uniqueness theorem for fractional differential equations, *J. Phys. A* **44**, pp. 392003.

Baleanu, D., Diethelm, K., Scalas, E. and Trujillo, J.J. (2012a). Fractional calculus models and numerical methods, Series on Complexity, Nonlinearity and Chaos, World Scientific, Boston.

Baleanu, D., Tenreiro Machado, J.A. and Luo, A. (2012b) Fractional Dynamics and Control, Springer.

Baleanu, D., Mustafa, O.G. and O'Regan, D. (2013). A uniqueness criterion for fractional differential equations with Caputo derivative, *Nonlinear Dyn.* **71**, pp. 635-640.

Baleanu, D. and Trujillo, J.J. (2008). On exact solutions of a class of fractional Euler-Lagrange equations, *Nonlin. Dyn.* **52**, pp. 331–335.

Baleanu, D. and Trujillo, J.J. (2010). A new method of finding the fractional Euler-Lagrange and Hamilton equations within Caputo fractional derivatives, *Comm. Nonlinear Sci.* **15**, pp. 1111–1115.

Baleanu, D., Mohamadi, H. and Rezapour, S. (2012c). Positive solutions of an initial value problem for nonlinear fractional differential equations, *Abstr. Appl. Anal.*, Article Number: 837437.

Baleanu, D., Mohamadi, H. and Rezapour, S. (2013a). The existence of solutions for a nonlinear mixed problem of singular fractional differential equations, *Adv. Diff. Equ.*, Article Number: 359.

Baleanu, D., Rezapour, S. and Mohamadi, H. (2013a).Some existence results on nonlinear fractional differential equations, *Phil. Trans. R. Soc. A*, **371**, Article Number: 20120144.

Baleanu, D., Nazemi, S.Z. and Rezapour, S. (2014a). The existence of solution for a k-dimensional system of multiterm fractional integrodifferential equations with antiperiodic boundary value problems, *Abstr. Appl. Anal.*, Article Number: 896871.

Baleanu, D., Nazemi, S.Z. and Rezapour, S. (2014b). A k-dimensional system of fractional neutral functional differential equations with bounded delay, *Abstr. Appl. Anal.*, Article Number: 524761.

Baleanu, D., Rezapour, S. and Saied, S. (2014). A k-dimensional system of fractional finite difference equations, *Abstr. Appl.. Anal..*, Article Number: 312578.

Bartušek, M., Došlá, Z. and Graef, J.R. (2004). The nonlinear limit–point/limit–circle problem, Birkhäuser, Boston.

Bashirov, A.E. and Mahmudov, N.I. (1999). On concepts of controllability for linear deterministic and stochastic systems. *SIAM J. Contr. Optim.*, **37**, pp. 1808–1821.

Belarbi, A., Benchohra, M. and Ouahab, A. (2006). Uniqueness results for fractional functional differential equations with infinite delay in Frechet spaces, *Applicable Anal.* **85**, pp. 1459–1470.

Bellman, R. (1953). Stability Theory of Differential Equations, McGraw-Hill, London.

Benchohra, M., Gorniewicz, L., Ntouyas, S.K. and Ouahab, A. (2004). Controllability results for impulsive functional differential inclusions, *Rep. Math. Phys.* **54**, pp. 211–228.

Bhrawy, A.H., Baleanu, D. and Assas, L.M. (2014). Efficient generalized Laguerrespectral methods for solving multi-term fractional differential equations on the half line, *J. Vib. Contr.* **20**, pp. 973–985.

Bhrawy, A.H. and Baleanu, D. (2013). A spectral Legendre-Gauss-Lobatto collocation method for a space-fractional advection diffusion equations with variable coefficients, *Rep. Math. Phys.* **72**, pp. 219–233.

Bian, W. (1998). Approximate controllability for semilinear systems. *Acta Math. Hung.* **81**, pp. 41–57.

Bihari, I. (1956). A generalization of a lemma of Bellman and its application to uniqueness problems of differential equations, *Acta Math. Acad. Sci. Hung.* **7**, pp. 81–94.

Bihari, I. (1957). Researches of the boundedness and stability of the solutions of nonlinear differential equations, *Acta Math. Acad. Sci. Hung.* **8**, pp. 260–278.

Brezis, H. (1999). Analyse fonctionelle. Théorie et applications, Dunod, Paris.

Burlak, J. (1965). On the nonexistence of L^2–solutions of a class of nonlinear differential equations, *Proc. Edinburgh Math. Soc.* **14**, pp. 257–268.

Byszewski, L. (1991). Theorems about the existence and uniqueness of solutions of a semilinear evolution nonlocal Cauchy problem, *J. Math. Anal. Appl.* **162**, pp. 494–505.

Byszewski, L. (1991). Theorems about the existence and uniqueness of continuous solutions of nonlocal problem for nonlinear hyperbolic equation, *Applicable Anal.* **40**, pp. 173–180.

Cao, J., Yang, Q., Huang, Z. and Liu, Q. (2011). Asymptotically almost periodic solutions of stochastic functional differential equations, *Appl. Math. Comput.* **218**, pp. 1499–1511.

Cao, J., Yang, Q. and Huang, Z. (2012). On almost periodic mild solutions for stochastic functional differential equations, *Nonlinear Anal. TMA* **13** (2012), pp. 275–286.

Chang, Y.K., Zhao, Z.H., N'Guérékata, G.M. and Ma, R. (2011). Stepanov-like almost automorphy for stochastic processes and applications to stochastic differential equations, *Nonlinear Anal. RWA* **12**, pp. 1130–1139.

Caponneto, R., Dongola, G., Fortuna, L. and Petras, I. (2010). Fractional Order Systems: Modeling and Control, World Scientific Series on Nonlinear Science Series A.

Chen, J., Xu, D.M. and Shafai, B. (1995). On sufficient conditions for stability independent of delay, *IEEE Transact. Autom. Contr.* **40**, pp. 1675–1680.

Chukwu, E.N. and Lenhart, S.M. (1991). Controllability question for nonlinear systems in abstract space, *J. Optim. Theory Appl.* **8**, pp. 437–462.

Constantin, A. (2010). On Nagumo's theorem, *Proc. Japan Acad.* **86(A)**, pp. 41–44.

Constantin, A. (1996). On the unicity of solutions for high-order differential equations, *Istituto Lombardo (Rend. Sci.) A* **130**, pp. 171–181.

Coppel, W.A. (1971). Disconjugacy, Lect. Notes Math. 220, Springer-Verlag, Berlin.

Corduneanu, C. (1973). Integral equations and stability of feedback systems, Academic Press, New York.

Daftardar-Gejji, V. and Babakhani, A. (2004). Analysis of a system of fractional differential equations, *J. Math. Anal. Appl.* **293**, pp. 511–522.

Daftardar-Gejji, V. (2005). Positive solutions of a system of non-autonomous fractional differential equations, *J. Math. Anal. Appl.* **302**, pp. 56–64.

Dannan, F.M. (1985). Integral inequalities of Gronwall-Bellman-Bihari type and asymptotic behavior of certain second order nonlinear differential equations, *J. Math. Anal. Appl.* **108**, pp. 151–164.

Dauer, J.P. and Mahmudov, V. (2002). Approximate controllability of semilinear functional equations in Hilbert spaces, *J. Math. Anal. Appl.* **273**, pp. 310–327.

Debbouche, A. and El-Borai, M.M. (2009). Weak almost periodic and optimal mild solutions of fractional evolution equations, *Electronic J. Diff. Eqs.* **46**, pp. 1–8.

Debbouche, A. (2010). Fractional evolution integro-differential systems with nonlocal conditions, *Adv. Dynam. Syst. Appl.* **5**, pp. 49–60.

Debbouche, A. and Baleanu, D. (2011). Controllability of fractional evolution nonlocal impulsive quasilinear delay integro-differential systems, *Comp. Math. Appl.* **62**, pp. 1442–1450.

Debbouche, A. and Baleanu, D. (2012). Exact null controllability for fractional nonlocal integrodifferential equations via implicit evolution system, *J. Appl. Math.*, **Volume 2012**, Article ID 931975, 17 pages, http://dx.doi.org/10.1155/2012/931975, pp. 1–17.

Debbouche, A., Baleanu, D. and Agarwal, R.P. (2012). Nonlocal nonlinear integro-differential equations of fractional orders, *Boundary Value Probl.*, **Volume 2012 78**, pp. 1–10.

Delbosco, D. and Rodino, L. (1996). Existence and uniqueness for a nonlinear fractional differential equation, *J. Math. Anal. Appl.* **204**, pp. 609–625.

Deng, K. (1993). Exponential decay of solutions of semilinear parabolic equations with nonlocal initial conditions, *J. Math. Anal. Appl.* **179**, pp. 630–637.

Deng, W., Li, C. and Lu, J. (2007). Stability analysis of linear fractional differential system with multiple time delays, *Nonlinear Dyn.* **48**, pp. 409–416.

Diethelm, K. and Ford, N.J. (2002). Analysis of fractional differential equations, *J. Math. Anal. Appl.* **265**, pp. 229–248.

Driver, R. (1977). Ordinary and delay differential equations, AMS 20, Springer, Berlin.

Doha, E.H., Bhrawy, A.H., Baleanu, D. and Hafez, R.M. (2014). A new Jacobi rational-Gauss collocation method for numerical solution of generalized pantograph equations, *Appl. Num. Math.* **77**, pp. 43–54.

Dugundji, J. and Granas, A. (1982). Fixed point theory I, Monogr. Matem. 61, PWN, Warszawa.

Eastham, M.S.P. (1970). Theory of ordinary differential equations, Van Nostrand Reinhold, London.

E.G. Edelman, M. and Tarasov, V.E. (2009). Fractional standard map, *Phys. Lett. A.* **374**, pp. 279–285.

El-Borai, M.M., El-Nadi, K.E. and El-Akabawy, E.G. (2010). On some fractional evolution equations, *Comp. Math. Appl.* **59**, pp. 1352–1355.

El-Borai, M.M. (2002). Some probability densities and fundamental solutions of fractional evolution equations, *Chaos Sol. Fractals* **14**, pp. 433–440.

El-Sayeed, M.A.A. (1996). Fractional order diffusion wave equation, *Int. J. Theor. Phys.* **35**, pp. 311–322.

El'sgol'tz, L.E. (1966). Introduction to the theory of differential equations with deviating arguments, Holden-Day, New York.

Evans, L.C. and Gariepy, R.F. (1992). Measure theory and fine properties of functions, CRC Press, Boca Raton.

Fite, W.B. (1918). Concerning the zeros of the solutions of certain differential equations, *Trans. Amer. Math. Soc.* **19**, pp. 341–352.

Furati, K.M. and Tatar, N.-e. (2005). Power-type estimates for a nonlinear fractional differential equation, *Nonlinear Anal. TMA* **62**, pp. 1025–1036.

Feckan, M., Wang, J.R. and Zhou, Y. (2013). Controllability of fractional functional evolution equations of Sobolev type via characteristic solution operators, *J. Optim. Theory. Appl.* **156**, pp. 79–95.

Gilbarg, D. and Trudinger N. (1998). Elliptic partial differential equations of second order, Springer-Verlag, Berlin.

Golomb, M. (1958). Bounds for solutions of nonlinear differential systems, *Arch. Rat. Mech. Anal.* **1**, pp. 272–282.

Graef, J.R. and Spikes, P.W. (1987). On the nonoscillation, convergence to zero, and integrability of solutions of a second order nonlinear differential equation, *Math. Nachr.* **130**, pp. 139–149.

Grammatikopoulos, M.K. and Kulenović, M.R. (1981). On the nonexistence of L^2–solutions of nth order differential equations, *Proc. Edinburgh Math. Soc.* **24**, pp. 131–136.

Hale, J.K. (1971). Functional differential equations, AMS 3, Springer-Verlag, New York.

Hale, J.K. and Verduyn Lunel, S.M. (1993). Introduction to functional-differential equations, AMS 99, Springer-Verlag, New York.

Hartman, P. and Wintner, A. (1955). On the assignment of asymptotic values for the solutions of linear differential equations of second order, *Amer. J. Math.* **77**, pp. 475–483.

Hartman, P. (1964). Ordinary differential equations, J. Wiley & Sons, New York.

Hille, E. (1959). Analytic function theory. Vol. I, Chelsea Publish., New York.

Hille, E. and Phillips, R.S. (1957). Functional analysis and semi-groups, Amer. Math. Soc., Coll. Publ. XXXI, Providence.

Ibragimov, N.H. and Kovalev, V.F. (2009). Approximate and renormgroup symmetries, Higher Education Press, Beijing and Springer-Verlag, Berlin.

Jafari, H., Haghbin, A., Hesam, S. and Baleanu, D. (2014). Solving partial q-differential equations within reduced q-differential transformation method, *Rom. J. Phys.* **59**, pp. 399–401.

Jarad, F., Abdeljawad, T., Baleanu, D. and Bicen, K. (2012). On the stability of some discrete fractional non autonomous systems, *Abstract Appl. Anal.* **2012**, Article ID 476581,http://dx.doi.org/10.1155/2012/476581.

Jeong, J.M., Kim, J.R. and Roh, H.H. (2007). Controllability for semilinear retarded control systems in Hilbert spaces, *J. Dynam. Control Syst.* **13**, pp. 577–591.

Jeong, J.M. and Kim, H.G. (2009). Controllability for semilinear functional integrodifferential equations, *Bull. Korean Math. Soc.* **46(3)**, pp. 463–475.

Jenson, V.G. and Jeffreys, G.V. (1997). Mathematical methods in chemical engineering. Second edition, Academic Press, New York.

Joshi, M.C. and Bose, R.K. (1985). Some topics in nonlinear functional analysis, John Wiley and Sons, New York.

Kartsatos, A.G. (1980). Advanced ordinary differential equations, Mariner Publ., Tampa, Florida.

Kauffman, R.M., Thomas, T. and Zettl, A. (1977). The deficiency index problem for powers of ordinary differential expressions, LNM 621, Springer-Verlag, Berlin.

Kequin, G., Kharitonov, V.L. and Chen, J. (2003). Stability of time-delay systems, Birkhäuser, Basel.

Kerboua, M., Debbouche, A. and Baleanu, D. (2013). Approximate controllability of Sobolev type nonlocal fractional stochastic dynamic systems in Hilbert spaces,*Abstr.Appl. Anal.* **Volume 2013**, Article ID 262191, 10 pages, http://dx.doi.org/10.1155/2013/262191.

Kiguradze, I.T. and Kvinikadze, G.G. (1982). On strongly increasing solutions of nonlinear ordinary differential equations, *Ann. Mat. Pura Appl.* **130**, pp. 67–87.

Kilbas, A.A. Srivastava, H.M., Trujillo, J.J. (2006). Theory and applications of fractional differential equations, North-Holland, New York.

Kiryakova, V. (1993). Generalized fractional calculus and applications, Longman and John Wiley, New York, USA.

Kharitonov, V.L. and Hinrichsen, D. (2004). Exponential estimates for time delay systems, *Syst. & Control Lett.* **53**, pp. 395–405.

Klimek, M. (2001). Fractional sequential mechanics - models with symmetric fractional derivative, *Czech. J. Phys.* **51**, pp. 1348–1354.

Klimek, M. (2009). On solutions of linear fractional differential equations of a variational type, The Publishing Office of Czestochowa University of Technology, Czestochowa, Poland.

Kolmogorov, A. and Fomine, S. (1974). Éléments de la théorie des fonctions et de l'analyse fonctionelle, Éd. Mir, Moscou (Moscow).

Krasovski, N.N. (1956). On the application of the second method of Lyapunov for equations with time retardations, *Prikl. Mat. Mek.* **20**, pp. 315–327.

Kumar, S. and Sukavanam, N. (2012). Approximate controllability of fractional order semilinear systems With bounded delay, *J. Diff. Equ.*, **252**, pp. 6163–6174.

Lakshmikantham, V. (2008). Theory of fractional functional differential equations, *Nonlinear Anal.* **69**, pp. 3337–3343.

Lakshmikantham, V., Bainov, D.D. and Simeonov, P.S. (1989). Theory of impulsive differential equations, World Scientific, Singapore.

Lakshmikantham, V. and Devi, J.V. (2008). Theory of fractional differential equations in Banach spaces, *European J. Pure Appl. Math.* **1**, pp. 38–45.

Lakshmikantham, V., Leela, S. and Devi, J.V. (2009). Theory of fractional dynamic systems, Cambridge Sci. Publish., Cottenham, UK.

Lakshmikantham, V. and Vatsala, A.S. (2008). Basic theory of fractional differential equations, *Nonlinear Anal. TMA* **69**, pp. 2677–2682.

Lakshmikantham, V. and Vatsala, A.S. (2008). General uniqueness and monotone iterative technique for fractional differential equations, *Appl. Math. Lett.* **21**, pp. 828–834.

Lang, S. (1999). Fundamentals of differential geometry, Springer-Verlag, New York.

Lanusse, P., Oustaloup, A. and Mathieu, B. (1993). Third generation CRONE control, in Proceedings of the IEEE International Conference on Systems, Man and Cybernetics *2*, pp. 149–155.

Lazarevic, M.P. (2006). Finite time stability analysis of fractional control of robotic time-delay systems, *Mech. Research Comm.* **33**, pp. 269–279.

Lee, J.W. and O'Regan, D. (1998). Topological transversality. Applications to initial value problems, *Ann. Polon. Math.* **48**, pp. 247–252.

Lee, T.N. and Dianat, S. (1981). Stability of time-delay systems, *IEEE Trans. Autom. Control* **26**, pp. 951–953.

Li, Y. Chen, Y.Q. and Podlubny, I. (2010). Stability of fractional-order nonlinear dynamic systems: Lyapunov direct method and generalized Mittag-Leffler stability, *Comp. Math. Appl.* **59**, pp. 1810–1821.

Li, F., Liang, J. and Xu, H.K. (2012). Existence of mild solutions for fractional integrodifferential equations of Sobolev type with nonlocal conditions, *J. Math. Anal. Appl.* **391**, pp. 510–525.

Lovelady, D.L. and Martin, Jr. R.H. (1972). A global existence theorem for a nonautonomous differential equation in a Banach space, *Proc. Amer. Math. Soc.* **35**, pp. 445–449.

Lovelady, D.L. (1973). A necessary and sufficient condition for exponentially bounded existence and uniqueness, *Bull. Austral. Math. Soc.* **8**, pp. 133–135.

Luchko, Y. and Kiryakova, V. (2013). The Mellin integral transform in fractional calculus, *Frac. Calc. Appl. Anal.* **16**, pp. 405–430.

Luo, Y. and Chen, Y.Q. (2013). Fractional Order Motion Control, John Wiley and Sons,Inc.

McShane, E.J. (1950). Linear functionals on certain Banach spaces. *Proc. Amer. Math. Soc.* **1**, pp. 402–408.

Mahmudov, N.I. (2003). Approximate controllability of semilinear deterministic and stochastic evolution equations in abstract spaces, *SIAM J. Contr. Opt.* **42**, pp. 1604–1622.

Mao, X. (1997). Stochastic Differential Equations and Applications, Horwood, Chichester.

Maraaba (Abdeljawad), T., Jarad, F. and Baleanu, D. (2008a). Existence and uniqueness theorem for a class of delay differential equations with left and right Caputo fractional derivatives, *J. Math. Phys.* **49**, pp. 083507.

Maraaba (Abdeljawad), T., Jarad, F. and Baleanu, D. (2008b). On the existence and the uniqueness theorem for fractional differential equations with bounded delay within Caputo derivatives, *Sci. China, Ser A: Math., Phys., Astron.* **51** (2008), pp. 1775–1786.

Merrikh-Bayat, F. and Karimi-Ghartemani, M. (2009). An efficient numerical algorithm for stability testing of fractional-delay systems, *ISA Transact.* **48**, pp. 32–37.

Metzler, R., Schick, W., Kilian, H.G. and Nonennmacher, T.F. (1995). Relaxation in filled polymers: A fractional calculus approach, *J. Chem. Phys.* **103**, pp. 7180–7186.

Miller, K.S. and Ross, B. (1993). An introduction to the fractional calculus and fractional differential equations, Wiley and Sons, New York.

Miller, K.S. and Ross, B. (1989). Fractional difference calculus, in Proceedings of the International Symposium on Univalent Functions, Fractional Calculus and Their Applications, Nihon University, Koriyama, Japan.

Momani, S. and Hadid, S. (2004). Lyapunov stability solutions of fractional integrodifferential equations, *Internat. J. Math. Math. Sci.* **45**, pp. 2503–2507.

Monje, A.C., Chen, Y.Q., Vinagre, B.M., Xue, D. and Feliu, V. (2010). Fractional-order Systems and Control (Fundametals and Applications), Springer.

Mophou, G.M and N'Guerekata, G.M. (2010). On some classes of almost automorphic functions and applications to fractional differential equations, *Comp. Math. Appl.* **59**, pp. 1310–1317.

Muldowney, J.S. and Wong, J.S.W. (1968). Bounds for solutions of nonlinear integro-differential equations, *J. Math. Anal. Appl.* **23**, pp. 487–499.

Mustafa, O.G. and Rogovchenko, Y.V. (2002). Global existence of solutions with prescribed asymptotic behavior for second-order nonlinear differential equations, *Nonlinear Anal. TMA* **51**, pp. 339–36.

Mustafa, O.G. and Rogovchenko, Y.V. (2003a), Limit–point type results for linear differential equations, *Arch. Inequal. Appl.* **1**, pp. 377–385.

Mustafa, O.G. and Rogovchenko, Y.V. (2003b). Existence of square integrable solutions of perturbed nonlinear differential equations, Proc. Fourth Internat. Conf. Dynam. Systems Diff. Eqs., Discrete Cont. Dynam. Syst., A supplement volume, pp. 647–655.

Mustafa, O.G. and Rogovchenko, Y.V. (2004a). Limit–point criteria for superlinear differential equations, *Bull. Belg. Math. Soc. Simon Stevin* **11**, pp. 431–440.

Mustafa, O.G. and Rogovchenko, Y.V. (2004b). Limit–point type solutions of nonlinear differential equations, *J. Math. Anal. Appl.* **294**, pp. 548–559.

Mustafa, O.G. (2005). On the existence interval in Peano's theorem, *Ann. A.I. Cuza Univ. Iaşi (Romania), Ser. Math., LI* (**1**), pp. 55–64.

Mustafa, O.G. (2007a). On some second order differential equations with convergent solutions, *Monatsh. Math. 150*, pp. 133–140.

Mustafa, O.G. (2007b). Hille's non-oscillation theorem and the decay of solutions to a class of semi-linear PDE's, *Arch. Math. (Basel)* **89**, pp. 452–458.

Mustafa, O.G. and Rogovchenko, Y.V. (2006). Asymptotic integration of a class of nonlinear differential equations, *Appl. Math. Lett.* **19**, pp. 849–853.

Mustafa, O.G. and Rogovchenko, Y.V. (2007). Estimates for domains of local invertibility of diffeomorphisms, *Proc. Amer. Math. Soc.* **135**, pp. 69–75.

Mustafa, O.G. (2008). Existence of positive evanescent solutions to some quasilinear elliptic equations, *Bull. Austral. Math. Soc.* **78**, pp. 157–162.

Mustafa, O.G. (2012). A Nagumo-like uniqueness result for a second order ODE, *Monatsh. Math.*, **168**, pp. 273–277, http://dx.doi.org/10.1007/s00605-011-0324-2.

Mustafa, O.G. and O'Regan, D. (2011). On the Nagumo uniqueness theorem, *Nonlinear Anal. TMA* **74**, pp. 6383–6386.

Nagumo, M. (1926). Eine hinreichende Bedingung für die Unität der Lösung von Differentialgleichungen erster Ordnung, *Japan J. Math.* **3**, pp. 107–112. Reprinted in: Mitio Nagumo Collected Papers, Eds. M. Yamaguti, L. Nirenberg, S. Mizohata, Y. Sibuya, Springer-Verlag, Tokyo, 1993.

Nagumo, M. (1927). Eine hinreichende Bedingung für die Unität der Lösung von gewöhnlichen Differentialgleichungen n-ter Ordnung, *Japan J. Math.* **4**, pp. 307–309. Reprinted in: Mitio Nagumo Collected Papers, Eds. M. Yamaguti, L. Nirenberg, S. Mizohata, Y. Sibuya, Springer-Verlag, Tokyo, 1993.

Nakagava, M. and Sorimachi, K. (1992). Basic characteristics of a fractance device, *IEICE Transact., Fundam.*, **E75-A**, pp. 1814–1818.

Nise, N.S. (2010). Contol Systems Engineering (Sixth Edition), John Wiley and Sons, Inc.

N'Guerekata, G.M. (2009). A Cauchy problem for some fractional abstract differential equation with non local conditions, *Nonlinear Anal. TMA* **70**, pp. 1873–1876.

Ogata, K. (2009). Modern Control Engineering (Fifth Edition), Prentice Hall.

Olver, F.J. (1974). Asymptotics and special functions, Academic Press, New York.

Oustaloup, A. (1991). *La Commande CRONE: Commande Robuste d'Ordre Non Entier* (Hermès, Paris).

Oustaloup, A., Cois, O., Lanusse, P., Melchior, P., Moreau, X. and Sabatier, J. (2006). The crone approach: Theoretical developments and major applications, in *Proceedings of the 2nd IFAC Workshop on Fractional Differentiation and its Applications* (Porto), pp. ID10.3182/20060719-3–PT–4902.00059.

Ouyang, L. (1957). The boundedness of solutions of linear differential equation $y'' + A(t)y = 0$, *Shuxue Jinzhan* **3**, pp. 409–415 (in Chinese).

Patula, W.T. and Waltman, P. (1974). Limit point classification of second order linear differential equations, *J. London Math. Soc.* **8**, pp. 209–216.

Pazy, A. (1983). Semigroups of linear operators and applications to partial differential equations, Springer-Verlag, New York.

Peano, G. (1890). Démonstration de l'intégrabilité des équations différentielles ordinaires, *Math. Ann.* **37**, pp. 182–228.

Pinto, M. (1998). Asymptotic integration of a system resulting from the perturbation of an h–system, *J. Math. Anal. Appl.* **131**, pp. 194–216.

Podlubny, I. (1999a). Fractional differential equations, Academic Press, San Diego.

Podlubny, I. (1999b). Fractional-order systems and $PI^\lambda D^\mu$-controllers, *IEEE Transact. Automat. Contr.* **44**, pp. 208–214.

Pucci, P. and Serrin, J. (1991). Continuation and limit properties for solutions of strongly nonlinear second order differential equations, *Asympt. Anal.* **4**, pp. 97–160.

Rabei, E.M., Nawafleh, K.I., Hijjawi, R.S., Muslih, S.I. and Baleanu, D. (2007). The Hamilton formalism with fractional derivatives, *J. Math. Anal. Appl.* **327**, pp. 891–897.

Razumikhin, B.S. (1956). On stability of systems with a delay, *Prikl. Matem. Mek.* **20**, pp. 500–512.

Raynaud, H.F. and Zergaınoh, A. (2000). State-space representation for fractional order controllers, *Automatica* **36**, pp. 1017–1021.

Richard, J.P. (2003). Time-delay systems: an overview of some recent advances and open problems, Automatica **39**, pp. 1667–1694.

Rogovchenko, Y.V. (1998). On the asymptotic behavior of solutions for a class of second order nonlinear differential equations, *Collect. Math.* **49**, pp. 113–120.

Rudin, W. (1987). Real and complex analysis. Third edition, McGraw-Hill, New York.

Sabatier, J. (2008). On stability of fractional order systems, in Proceedings of the Plenary Lecture VIII on 3rd IFAC Workshop on Fractional Differentiation and Its Applications, Ankara, Turkey.

Sadati, S.J., Baleanu, D., Ranjbar, A., Ghaderi, R. and Abdeljawad (Maraaba), T. (2010). Mittag-Leffler stability theorem for fractional nonlinear systems with delay, *Abstract Appl. Anal. 2010*, Article ID 108651, http://dx.doi.org/10.1155/2010/108651.

Sakthivel, R., Choi, Q.H. and Anthoni, S.M. (2004). Controllability result for nonlinear evolution integrodifferential systems, *Appl. Math. Lett.* **17**, pp. 1015–1023.

Sakthivel, R., Anthoni, S.M. and Kim, J.H. (2005). Existence and controllability result for semilinear evolution integrodifferential systems, *Math. Comp. Model.* **41**, pp. 1005–1011.

Sakthivel, R. and Anandhi, E.R. (2009). Approximate controllability of impulsive differential equations with state-dependent delay, *Intern. J. Control* **83**, pp. 387–393.

Sakthivel, R., Mahmudov, N.I. and Nieto, J.J. (2012a). Controllability for a class of fractional-order neutral evolution control systems, *Appl. Math. Comput.* **218**, pp. 10334–10340.

Sakthivel, R. Suganya, S. and Anthoni, S.M. (2012b). Approximate controllability of fractional stochastic evolution equations, *Comput. Math. Appl.* **63**, pp. 660–668.

Sakthivel, R. and Ren, Y. (2012). Approximate Controllability of Fractional Differential Equations with State-Dependent Delay, *Results in Mathematics*, pp. 1–15.

Sakthivel, R., Ren, Y. and Mahmudov, N.I. (2011). On the approximate controllability of semilinear fractional differential systems, *Comput. Math. Appl.*, **62**, pp. 1451–1459.

Sakthivel, R., Revathi, P. and Renc, Y. (2013). Existence of solutions for nonlinear fractional stochastic differential equations, *Nonlinear Analysis*, **81**, pp. 70–86.

Sukavanam, N. and Kumar, S. (2011). Approximate controllability of fractional order semilinear delay systems, *J. Optim. Theory Appl.*, **151** pp. 373–384.

Samko, S.G., Kilbas, AA. and Marichev, O.I. (1993). Fractional integrals and derivatives. Theory and applications, Gordon and Breach, Switzerland.

Sign, C. and Chuanzhi, B. (2006). Positive solutions for nonlinear fractional differential equations with coefficient that changes sign, *Nonlinear Anal. TMA* **64**, pp. 677–685.

Srivastava, H.M., Golmankhaneh, A.K., Baleanu, D. and Yang, X.J. (2014). Local fractional Sumudu transform with application to IVPs on Cantor sets, *Abstr. Appl.. Anal.*. Article Number: 620529.

Subalakshmi, R. and Balachandran, K. (2009). Approximate controllability of nonlinear stochastic impulsive integrodifferential systems in Hilbert spaces, *Chaos Sol. Fractals* **42**, pp. 2035–2046.

Stein, E.M. (1970). Topics in harmonic analysis. Related to the Littlewood-Paley theory, AMS 63, Princeton Univ. Press, New Jersey.

Sukavanam, N. and Kumar, S. (2011). Approximate controllability of fractional order semilinear delay systems, *J. Optim. Theory Appl.*, **151**, pp. 373–384.

Sukavanam, N. and Tomar, N.K. (2007). Approximate controllability of semilinear delay control systems, *Nonlinear Funct. Anal. Appl.* **12**, pp. 53–59.

Tatar, N.-E (2010). On a boundary controller of fractional type, *Nonlinear Anal.* **72**, pp. 3209–3215.

Tai, Z. and Wang, X. (2009). Controllability of fractional-order impulsive neutral functional infinite delay integrodifferential systems in Banach spaces, *Appl. Math. Lett.* **22**, pp. 1760–1765.

Torvik, P.J. and Bagley, R.L. (1984). On the appearance of the fractional derivative in the behavior of real materials, *J. Appl. Mech.* **51**, pp. 294–298.

Trench, W.F. (1963). On the asymptotic behavior of solutions of second order linear differential equations, *Proc. Amer. Math. Soc.* **14**, pp. 12–14.

Triggiani, R. (1977). A note on the lack of exact controllability for mild solutions in Banach spaces, *SIAM J. Control Optim.* **15**, pp. 407–41.

Valerio, D., Trujillo, J.J., Rivero, M., Machado, J.A.T. and Baleanu, D. (2013). Fractional calculus: A survey of useful formulas, *Eur. J. Phys.* **222**, pp. 1827–1846.

Wang, J. R. and Y. Zhou, Y. (2011). A class of fractional evolution equations and optimal controls, *Nonlinear Anal., Real World Appl.*, **12**, pp. 262–272.

Walter, W. (1998). Ordinary differential equations, Springer-Verlag, New York.

Waltman, P. (1964). On the asymptotic behavior of solutions of a nonlinear equation, *Proc. Amer. Math. Soc.* **15**, pp. 918–923.

Weber, E. (1956). Linear transient analysis, vol. II, John Wiley & Sons, New York.

Weyl, H. (1910). Über gewöhnliche differentialgleichungen mit singularitäten und die zugehörige entwocklungen willkülricher funktionen, *Math. Ann.* **68**, pp. 220–269.

Whyburn, G.T. (1964). Topological analysis. Revised edition, Princeton Univ. Press, New Jersey.

Wintner, A. (1935). On the exact limit of the bound for the regularity of solutions of ordinary differential equations, *Amer. J. Math.* **57**, pp. 539–540.

Wintner, A. (1947). Asymptotic integrations of the adiabatic oscillator, *Amer. J. Math.* **69**, pp. 251–272.

Wintner, A. (1950). A criterion for the nonexistence of L^2–solutions of a nonoscillatory differential equation, *J. London Math. Soc.* **25**, pp. 347–351.

Wintner, A. (1956). On the local uniqueness of the initial value problem of the differential equation $\frac{d^n x}{d^n t} = f(t, x)$, *Boll. Un. Mat. Ital.* **11**, pp. 496–498.

Wu, G.C. and Baleanu, D. (2014a). Discrete fractional logistic map and its chaos, *Nonlinear Dyn.* **75**, pp. 283–287.

Wu, G.C. Baleanu, D. and Zeng, S.D. (2014). Discrete Chaos in Fractional Sine and Standard Maps, *Phys. Lett. A* **378**, pp. 484–487.

Wu, G.C. and Baleanu, D. (2014b). Discrete chaos in fractional delayed logistic maps, *Nonlinear Dyn.* **75**, pp. 283–287.

Yan, Z. (2012a). Approximate controllability of fractional neutral integro-differential inclusions with state-dependent delay in Hilbert spaces, *IMA J. Math. Contr. Infor.* **12**, pp. 1–20.

Yan, Z. (2012b). Approximate controllability of partial neutral functional differential systems of fractional order with state-dependent delay, *Int. J. Contr.*, **85**, pp. 1051–1062.

Ye, H., Ding, Y. and Gao, J. (2007). The existence of a positive solution of a fractional differential equation with delay, *Positivity* **11**, pp. 341–350.

Ye, H.P. and Gao, J.M. (2011). Henry-Gronwall type retarded integral inequalities and their applications to fractional differential equations with delay, *Appl. Math. Comp.* **218**, pp. 4152–4160.

Zhang, X. (2008). Some results of linear fractional order time-delay system, *Appl. Math. Comp.* **197**, pp. 407–411.

Zhou, Y. and Jiao, F. (2010a). Nonlocal Cauchy problem for fractional evolution equations, *Nonlinear Anal., Real World Appl.*, **11**, pp. 4465–4475

Zhou, Y. and Jiao, F. (2010b). Existence of mild solutions for fractional neutral evolution equations, *Comput. Math. Appl.*, **59**, pp. 1063–1077.

Index

A-resolvent family, 158
C_0-semigroup, 172
C_r–condition, 125

The right fractional sum, 140

Abel series, 10
approximate controllability, 179
Arzelà-Ascoli Theorem, 136
asymptotically linear solutions, 91

Banach space, 35
Beta function, 2
Bihari inequality, 57, 58

Caputo differential operator, 22, 27, 123
control function, 161
Controllability, 157

Discrete Fractional Nonautonomous Systems, 140
dual Bihari inequality, 60

Emden-Fowler, 95
Euler integral of first kind, 2
exponential stability, 147

Fite length criterion, 53
fractional differential equations with delay, 129

fractional nonlocal impulsive integro-differential control system, 161
Fubini theorem, 7

Gamma function, 2
globally uniformly asymptotically stable, 141

Hölder's inequality, 15, 62, 174
Hilbert–Schmidt operators, 172

identically null solution, 43, 44
intermediate asymptotic, 91

Lebesgue integral, 33
Lebesgue measure, 3
Lipschitz-type formula, 73
Lovelady-Martin uniqueness theorem, 35
Lyapunov function, 141, 151

mild solution, 41, 43, 44, 53
Mittag-Leffler function, 146
Mittag-Leffler stability, 145

Nagumo theorem, 40
Nagumo uniqueness criterion, 44
nonlocal fractional stochastic system, 170

Ouyang's inequality, 60

Peano existence theorem, 45

Razumikhin stability theory, 151

Riemann-Liouville, 2

Riemann-Liouville derivative, 3

sigma algebra, 171

space of continuous functions, 130

strongly measurable mapping, 172

Trench-type asymptotic, 83

uniform stability, 152

uniformly asymptotically stable, 141